NATIONAL 5

BIOLOGY

SECOND EDITION

James Torrance with
Caroline Stevenson, Clare Marsh,
James Fullarton & James Simms

**HODDER
GIBSON**
AN HACHETTE UK COMPANY

The Publishers would like to thank the following for permission to reproduce copyright material:

Photo credits

p. 1 (background) and Section 1 running head image © beawolf / stock.adobe.com; figure 1.4 © Lester V. Bergman / Corbis Documentary via Getty Images; figure 2.16 © Biophoto Associates / SPL; figure 3.3 © Science Source / SPL; figure 4.1 © Prof. K.Seddon & Dr. T.Evans, Queen's University Belfast / SPL; figure 4.2 © MOLEKUUL / SCIENCE PHOTO LIBRARY / Getty Images; figure 5.2 © Nigel Cattlin / Alamy Stock Photo; figure 6.6 © CNRI/SPL; figure 6.11 © John Fryer / Alamy; p. 49 (background) and Section 2 running head image © V. Yakobchuk / stock.adobe. com; figure 7.3 © Power and Syred / SPL; figure 7.12 © Professor Miodrag Stojkovic / SPL; figure 7.15 © Avita Medical; figure 7.16 © peretzp, https://www.flickr.com/photos/ifl/6327625030/in/album-72157627961304009/ (https://creativecommons.org/licenses/by-sa/2.0/); figure 8.3 © Dr Colin Chumbley / SPL; figure 8.5 © CNRI / SPL; figure 8.10a © Medicshots / Alamy Stock Photo; figure 8.10b © Chris Rout / Alamy Stock Photo; figure 9.3 © Innerspace Imaging / SPL; figure 9.4 © MedicalRF.com via Getty Images; figure 10.19 © Martyn Chillmaid / SPL; figure 12.4 © Nathan Benn / Corbis via Getty Images; figure 12.8 © Lunagrafix / SPL; figure 12.9 © Steve Gschmeissner / SPL; figure 13.7 © Biophoto Associates / SPL; p. 115 (background) and Section 3 running head image © Frank Boston / stock.adobe.com; figure 14.8 © blickwinkel / Alamy; figure 14.9 © davidyoung11111 / stock.adobe.com; figure 14.10 © david tipling / Alamy; figure 14.11 © David Whitaker / Alamy; figure 14.13a © hfox / stock.adobe.com; figure 14.13b © Shelli Jensen / stock.adobe.com; figure 14.14 © Michael Durham / Minden Pictures / Getty Images; figure 14.19 © giedriius / stock.adobe.com; figure 18.2 © Elenathewise / stock.adobe.com; figure 18.3 © Dario Sabljak / Alamy; figure 18.5 © Daniel Poloha / stock.adobe.com; figure 18.9 © lamax / stock.adobe.com; figure 18.15 © FLPA / Alamy; figure 18.16 © Wayne Lawler/ SPL; figure 19.3 © RAGUET/PHANIE/REX/Shutterstock; figure 19.4 © Hattie Young/SPL; figure 19.8 © Hakan Soderholm / Alamy Stock Photo; figure 19.9 © Bazzano Photography / Alamy Stock Photo; figure 19.13 © Biomedical Imaging Unit / Southampton General Hospital / SPL; figure 19.14 © PEGGY GREB / US DEPARTMENT OF AGRICULTURE / SCIENCE PHOTO LIBRARY; figure 19.18a © David Hosking / Alamy Stock Photo; figure 19.18b © Andrew Tiley / Alamy.

All other photos supplied by James Torrance.

Every effort has been made to trace all copyright holders, but if any have been inadvertently overlooked the Publishers will be pleased to make the necessary arrangements at the first opportunity.

While every effort has been made to check the instructions of practical work in this book, it is still the duty and legal obligation of schools to carry out their own risk assessments.

Although every effort has been made to ensure that website addresses are correct at time of going to press, Hodder Gibson cannot be held responsible for the content of any website mentioned in this book. It is sometimes possible to find a relocated web page by typing in the address of the home page for a website in the URL window of your browser.

Hachette UK's policy is to use papers that are natural, renewable and recyclable products and made from wood grown in sustainable forests. The logging and manufacturing processes are expected to conform to the environmental regulations of the country of origin.

Orders: please contact Hachette UK Distribution, Hely Hutchinson Centre, Milton Road, Didcot, Oxfordshire, OX11 7HH. Telephone: +44 (0)1235 827827. Email education@hachette.co.uk. Lines are open from 9 a.m. to 5 p.m., Monday to Friday. You can also order through our website: www.hoddereducation.co.uk. If you have queries or questions that aren't about an order, you can contact us at hoddergibson@hodder.co.uk

Cover photo © Brandon Alms/123RF.com

Illustrations by James Torrance

Typeset in Minion Regular 11/14 by Integra Software Services Pvt. Ltd., Pondicherry, India

Printed and bound by CPI Group (UK) Ltd, Croydon, CR0 4YY

A catalogue record for this title is available from the British Library

SCOTLAND EXCEL

We are an approved supplier on the Scotland Excel framework.

Schools can find us on their procurement system as:

Hodder & Stoughton Limited t/a Hodder Gibson.

MIX
Paper from responsible sources
FSC
www.fsc.org
FSC™ C104740

Contents

Preface

This book is designed to be a valuable resource for students studying SQA National 5 Biology. Each section of the book matches a section of the syllabus. Each chapter corresponds to a syllabus sub-topic. The text is presented in a format that allows clear differentiation between the mandatory core text and non-mandatory learning activities. These take the form of *Related Topics, Related Activities, Case Studies, Research Topics* and *Investigations.*

Each chapter includes one or two sets of *Testing Your Knowledge* questions to allow knowledge of core content to be assessed continuously throughout the course and to check that full understanding has been achieved.

At regular intervals throughout the book, *What You Should Know* summaries of key facts and concepts are given as 'cloze' tests accompanied by appropriate word banks. These provide an excellent source of material for consolidation and revision.

Each section ends with a varied selection of *Applying Your Knowledge and Skills* questions to foster the development of the mandatory subject skills outlined in the course arrangements. The questions are designed to prepare students for National 5 assessment, where they will be expected to demonstrate their ability to solve problems, select relevant information, present information, process data, plan experimental procedures, evaluate experimental designs, draw valid conclusions and make predictions and generalisations.

Further problem-solving exercises are available in the associated book *National 5 Biology: Applying Knowledge and Skills.*

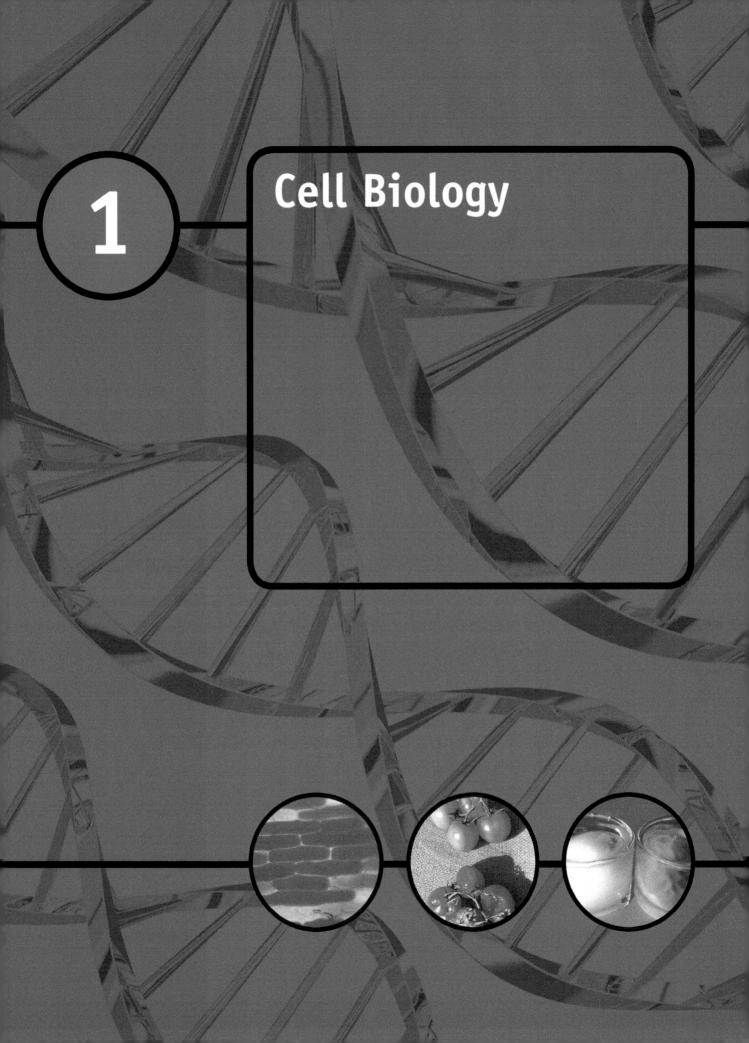

1 Cell Biology

Cell structure

Cells

Every living organism is made up of one or more **cells**. Cells are the basic units of life. Nothing smaller than a cell can lead an independent life and show all the characteristics of a living thing.

Figures 1.1 and 1.2 show examples of plant cells viewed under the microscope. Some of a plant cell's parts can be clearly seen when the cell is mounted in water. An *Elodea* leaf cell, for example, is seen to possess a **cell wall** and several green **chloroplasts**. Other cell structures that are not so obvious can often be shown up more clearly by the addition of dyes called stains. Iodine solution, for example, can be used to stain **nuclei**, as shown in Figure 1.2.

Figure 1.1 *Elodea* leaf cells in water

Close examination of the four types of cell shown in Figure 1.3 reveals that they have some features in common but also differ from one another in several ways. The functions of their various cellular structures are summarised in Table 1.1 on page 4.

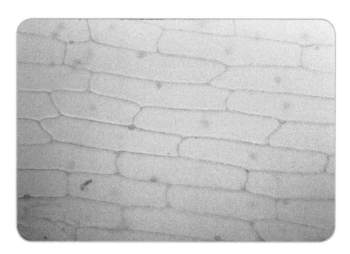

Figure 1.2 Onion leaf cells in iodine solution

Organelles are tiny structures (such as chloroplasts) that are:

- present in a cell's cytoplasm as discrete units, normally surrounded by a membrane
- responsible for a specialised function (such as photosynthesis).

A bacterium does not possess membrane-bound organelles. For example, it lacks a **nucleus** containing **threadlike chromosomes**. Instead, its genetic material takes the form of a single **circular chromosome** and several smaller circular **plasmids**. The structure and composition of the **cell wall** of a bacterium, a fungal cell and a green plant cell are all different from one another. Only plant cells have walls composed of **cellulose** (also see Figure 16.3 on page 143).

Ultrastructure

Tiny structures such as mitochondria and ribosomes form part of a cell's **ultrastructure**. Such minute structures can only be seen in detail using an electron microscope. Figure 1.4 shows an **electron micrograph** (a photo of the image produced using an electron microscope).

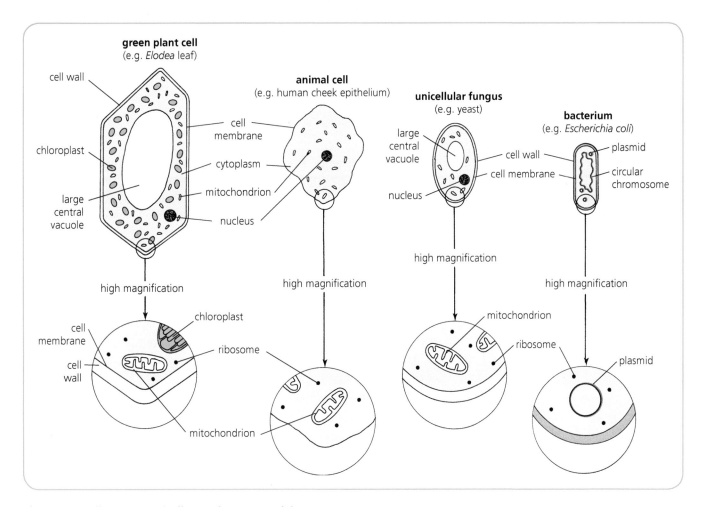

Figure 1.3 Cell structures (cells not drawn to scale)

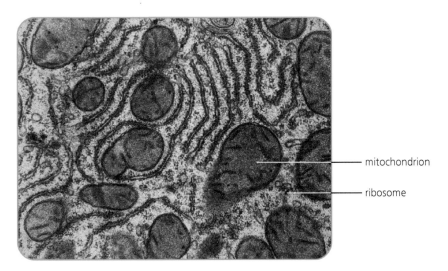

Figure 1.4 Electron micrograph showing mitochondria and ribosomes

Location	Cell structure	Description	Function
found in cells of green plants, animals, fungi and bacteria	cytoplasm	fluid or jellylike material containing essential materials	acts as a medium within which many important biochemical reactions take place
	cell membrane	thin layer surrounding cytoplasm	controls passage of substances into and out of cell (see Chapter 2)
found in cells of green plants, animals and fungi	nucleus	large, normally spherical structure containing threadlike chromosomes	controls cell's activities and passes information on from cell to cell (see Chapter 7) and from generation to generation (see Chapter 10)
	mitochondrion (plural mitochondria)	one of many tiny, sausage-shaped structures containing enzymes	aerobic respiration (see Chapter 6)
found in cells of green plants and fungi	large central vacuole	fluid-filled sac-like structure in cytoplasm	stores water and solutes as cell sap and regulates water content by osmosis (see Chapter 2)
found in cells of green plants, fungi and bacteria	cell wall	outer layer made of basket-like mesh of fibres (structure and composition of cell walls of fungi and bacteria differ from that of green plant cells)	supports the cell (see Chapter 2)
found in cells of green plants only	chloroplast	one of many discus-shaped structures containing green chlorophyll	photosynthesis (see Chapter 16)
found in cells of bacteria	circular chromosome	single ring of genetic material	control of cell activities and transfer of genetic information from cell to cell (see Chapter 5)
	plasmid	one of several tiny rings of genetic material	
found in all cells	ribosome	one of many tiny particles lacking membrane boundary	protein synthesis (see Chapter 3)

Table 1.1 Functions of cell structures

Related Activity

Measuring cell size

When a small piece of graph paper with several pinholes, 1 mm apart, is viewed under the low-power lens of a microscope, the pinholes are easily spotted because light passes up through them. Therefore the diameter of the microscope's field of view can be estimated. Figure 1.5 a), for example, shows a field of view with a diameter of 2 mm. Since **1 millimetre (mm) = 1000 micrometres (μm)**, the diameter of the field of view for this microscope lens = 2000 μm.

Next a sample of cells is viewed and the average number of cells lying end to end along the diameter of the field of view is found. Figure 1.5 b), for example, shows this number to be ten for a specimen of rhubarb epidermal cells. Since the length of ten cells = the diameter of the field of view = 2000 μm, the average length of one rhubarb epidermal cell = 2000/10 = 200 μm. Similarly, since the breadth of 20 cells = 2000 μm, the average breadth of one rhubarb epidermal cell = 2000/20 = 100 μm.

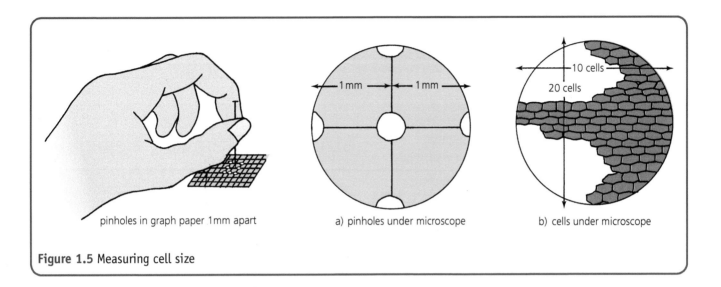

pinholes in graph paper 1mm apart

a) pinholes under microscope

b) cells under microscope

Figure 1.5 Measuring cell size

Testing Your Knowledge

1 What name is given to the basic units of life that can lead an independent existence? (1)

2 a) Name FOUR structural features that a typical plant cell and a typical animal cell have in common. (4)

   b) Identify THREE structural features present in an *Elodea* leaf cell but absent from a cheek epithelial cell. (3)

3 Give the function of each of the following structures: chloroplast, nucleus, cell membrane, mitochondrion. (4)

4 a) Express 1 millimetre in micrometres. (1)

   b) Express 1 micrometre as a decimal fraction of a millimetre. (1)

2 Transport across cell membranes

In a plant cell, the cell membrane enclosing the cytoplasm lies against the inside of the cell wall. The large central vacuole and other organelles are each surrounded by a membrane similar in structure to that of the cell membrane.

Structure of the cell membrane

The cell membrane is now known to be made up of **protein** and **phospholipid** molecules. It is thought to consist of a double layer of phospholipid molecules

Investigation

Structure of cell membranes

The cell sap present in the central vacuole of a beetroot cell (see Figure 2.1) contains red pigment. 'Bleeding' (the escape of this red sap from a cell) indicates that the cell membrane and vacuolar membrane have been damaged.

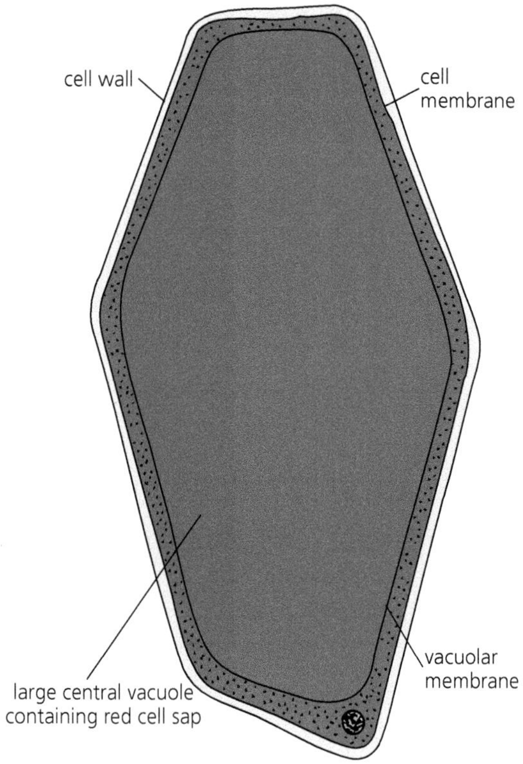

cell wall

cell membrane

vacuolar membrane

large central vacuole containing red cell sap

Figure 2.1 Beetroot cell

In the experiment shown in Figure 2.2, four identical cylinders of fresh beetroot are prepared using a cork borer. The cylinders are thoroughly washed in distilled water to remove traces of red cell sap from outer damaged cells.

The figure shows the results of subjecting the cylinders to various conditions. Bleeding is found to occur in B, C and D showing that the membranes have been destroyed.

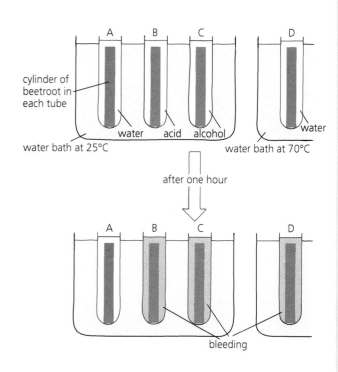

cylinder of beetroot in each tube

water   acid   alcohol   water

water bath at 25°C   water bath at 70°C

after one hour

bleeding

Figure 2.2 Investigating the chemical nature of the cell membrane

When molecules of protein are exposed to acid or high temperatures, they are known to become denatured (destroyed and non-functional). Molecules of lipid are known to be soluble in alcohol. It is therefore concluded that the cell membrane contains **protein** (as indicated by B and D whose denatured protein has allowed red cell sap to leak out) and **lipid** (as indicated by C whose lipid molecules have dissolved in alcohol, permitting the pigment to escape).

containing a patchy mosaic of protein molecules. This is known as the **fluid mosaic model** (see Figure 2.3).

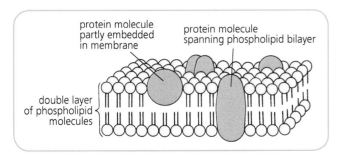

Figure 2.3 Fluid mosaic model of cell membrane

Diffusion

The molecules of a liquid (or a gas) move about freely all the time. In the experiment shown in Figure 2.4, a crystal of purple potassium permanganate is dropped into a beaker of water. The diagram illustrates the events that follow. The purple particles move from a region of **high concentration** (the dissolving crystal) to a region of **low concentration** (the surrounding water) until the concentration of purple particles (and water) is uniform throughout.

Diffusion is the name given to the movement of the molecules of a substance from a region of high concentration of that substance to a region of low concentration until the concentration becomes equal.

Concentration gradient

The difference in concentration that exists between a region of high concentration and a region of low concentration is called the **concentration gradient**. During diffusion, movement of molecules always occurs down a concentration gradient from high to low concentration. Diffusion is a **passive** process. This means that it does not require energy.

Importance of diffusion in cells

Unicellular organisms

Diffusion of oxygen and carbon dioxide

Different concentrations of substances exist inside a cell compared with outside in its environment. In a unicellular animal such as *Paramecium*, oxygen is constantly being used up by the cell contents during respiration. This results in the concentration of oxygen molecules inside the cell being lower than in the surrounding water. The cell membrane is freely permeable to the tiny oxygen molecules. Oxygen therefore **diffuses into** the cell from a higher concentration to a lower concentration (see Figure 2.5).

At the same time, the living cell contents are constantly making carbon dioxide by respiration. This results in the concentration of carbon dioxide inside the cell being higher than in the surrounding water.

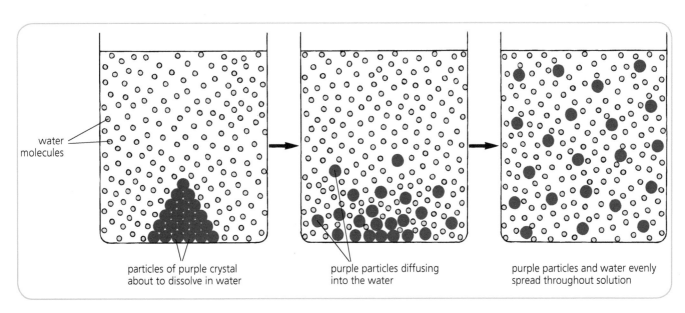

Figure 2.4 Diffusion

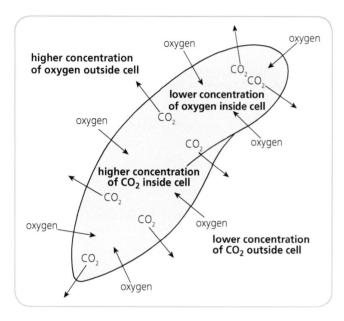

Figure 2.5 Diffusion into and out of a cell

Multicellular animals

In a multicellular animal such as a human being, diffusion also plays an important role in the exchange of respiratory gases. Blood returning to the lungs from respiring cells (see Figure 2.6) contains a higher concentration of carbon dioxide and a lower concentration of oxygen than the air in the air sac. Carbon dioxide therefore **diffuses out** of the blood and oxygen **diffuses in**. When the blood reaches living body cells, the reverse process occurs and the cells gain oxygen from the blood and lose carbon dioxide by diffusion. Similarly, diffusion is essential for the movement, down a concentration gradient, of molecules of dissolved food (such as glucose and amino acids) from the animal's bloodstream to its respiring cells.

Since the cell membrane is also freely permeable to tiny carbon dioxide molecules, these **diffuse out**, as shown in Figure 2.5.

Role of the cell membrane

Although the membrane of a cell is freely permeable to small molecules such as oxygen, carbon dioxide and water, it is not equally permeable to all substances.

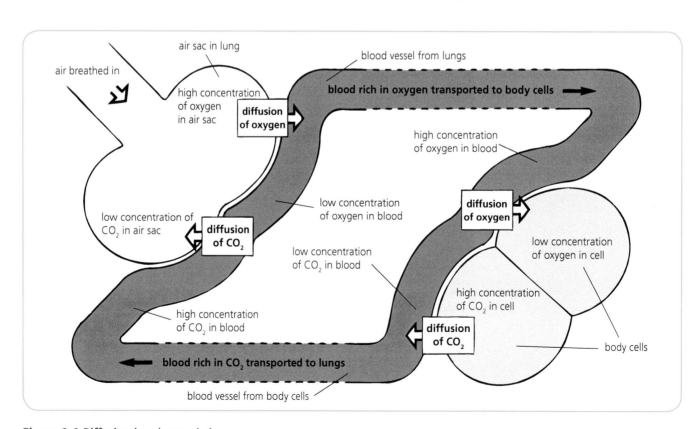

Figure 2.6 Diffusion in a human being

1 a) Name TWO types of molecule present in the cell membrane. (2)
  b) Which of these would be destroyed (denatured) at 70 °C? (1)
2 Define the terms *diffusion* and *concentration gradient*. (4)
3 a) Name TWO essential substances that enter an animal cell by diffusion. (2)
  b) Name a waste material that diffuses out of an animal cell. (1)
4 a) With reference to Figure 2.6, explain why diffusion is important to human beings. (2)
  b) i) Predict what would happen to the rate of diffusion of CO_2 if a person exercised vigorously.
    ii) Explain your answer. (3)

Larger molecules such as dissolved food can pass through the membrane slowly. Molecules that are even larger, such as starch, are unable to pass through. Thus the cell membrane controls the passage of substances into and out of a cell.

The exact means by which the cell membrane exerts this control is not fully understood. It is known that most cell membranes possess **tiny pores**. It is thought that many small molecules enter or leave by these pores. Other molecules are kept inside or outside the cell because they are too big to pass through the pores

(see Figure 2.7). The cell membrane is therefore said to be **selectively permeable**.

Osmosis

In both of the investigations shown on page 10, overall movement of tiny water molecules occurred from a region of **higher water concentration (HWC)** to a region of **lower water concentration (LWC)** across a membrane. The larger sugar molecules were unable to diffuse through the membrane because the membrane was **selectively permeable**. This movement of water molecules from a HWC to a LWC through a selectively permeable membrane is a special case of diffusion called **osmosis**. It is a **passive** process.

Water concentration gradient

The difference in water concentration that exists between two regions is called the **water concentration gradient**. At the start of the experiment shown in Figure 2.11 on page 11, a water concentration gradient exists between the regions on either side of the Visking tubing membrane in cell models A and C. During osmosis, water molecules move down a water concentration gradient from high to low water concentration through the selectively permeable membrane.

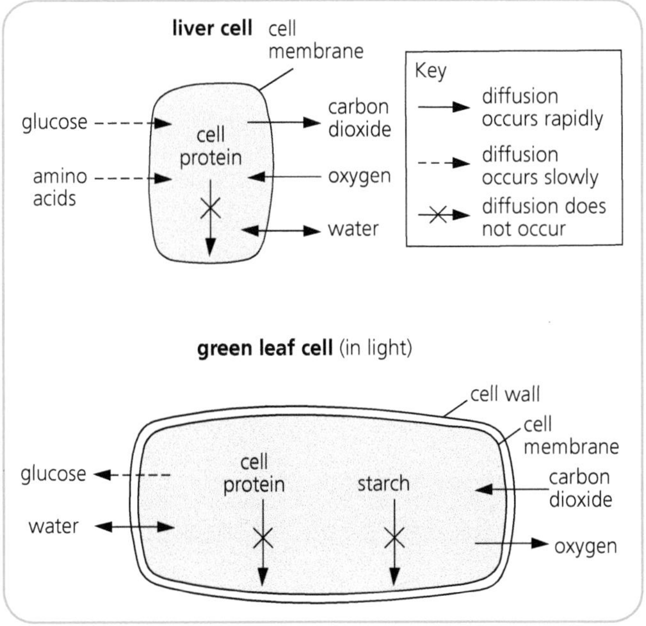

Figure 2.7 Diffusion into and out of different cells

Effect of water and concentrated sugar solution on potato cylinders

In the experiment shown in Figure 2.8, the very dilute sugar solution outside potato cylinder A has a higher water concentration (HWC) than the contents of the potato cells, which have a lower water concentration (LWC). The contents of the potato cells in cylinder B have a higher water concentration (HWC) than the surrounding sugar solution, which has a lower water concentration (LWC).

After 24 hours, cylinder A is found to have increased in volume and mass and to have become firmer (turgid) in texture. Cylinder B, on the other hand, has decreased in both volume and mass and has become softer (flaccid). It is therefore concluded that water molecules have diffused into cylinder A and out of cylinder B through the cell membranes.

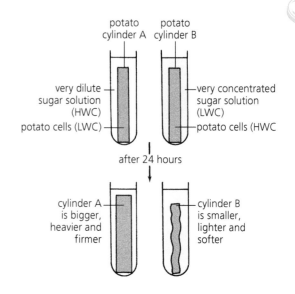

Figure 2.8 Effect of two sugar solutions on potato cells

Effect of water and concentrated sugar solution (syrup) on eggs

In the experiment shown in Figures 2.9 and 2.10, each egg is surrounded by its membrane only. Its shell has been removed by acid treatment prior to the experiment. After two days, egg A swells up whereas egg B shrinks. It is therefore concluded that water has diffused into egg A (from a higher water concentration to a lower water concentration) and out of egg B (from a higher water concentration to a lower water concentration).

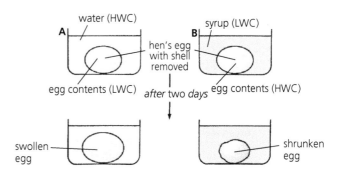

Figure 2.10 Osmosis in eggs

Figure 2.9 Osmosis takes effect

Related Activity

Constructing cell models of osmosis

Visking tubing is a synthetic material. Over a short time-scale (such as two hours) it acts like the selectively permeable membrane that surrounds a cell. Lengths of Visking tubing tied at both ends can therefore be used to act as models of cells.

Figure 2.11 shows three cell models (A, B and C) set up to investigate osmosis. Cell model A is found to gain mass since water has passed into the 'cell' from a region of higher water concentration (HWC) to a region of lower water concentration (LWC) by osmosis. B neither gains nor loses water since the solutions inside and outside are equal in water concentration. C is found to lose mass since water has passed out of the 'cell' from a region of HWC to a region of LWC by osmosis.

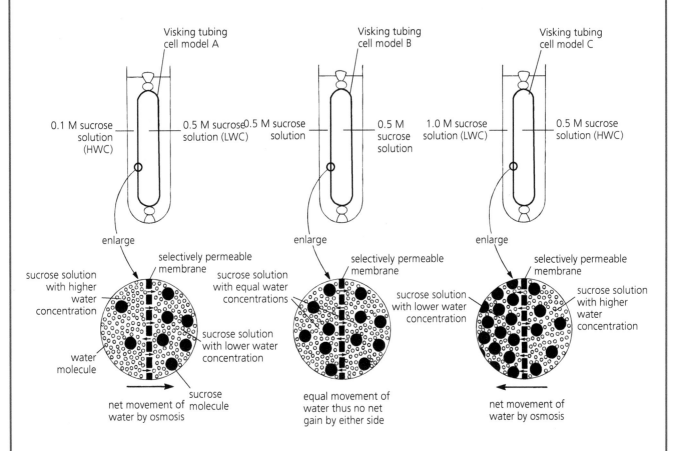

Figure 2.11 Molecular model of osmosis (HWC = region of high water concentration, LWC = region of low water concentration)

Osmosis and cells

Movement of water by osmosis occurs between cells and their immediate environment. The direction depends on the water concentration of the liquid in which the cells are immersed compared with that of the cell contents.

Red blood cells

Since pure water has a higher water concentration than the contents of red blood cells, water enters by osmosis until the cells burst (see Figure 2.12). Since 0.85% salt solution has the same water concentration as the cell contents, there is no net flow of water into or out of the cells by osmosis (see Figures 2.12 and 2.13). Since 1.7% salt solution has a lower water concentration than the cells, water passes out and the cells shrink (see Figures 2.12 and 2.14).

Plant cells

Plant cells such as onion cells mounted in liquids of different water concentrations respond in different ways. Pure water has a higher water concentration than the contents of a normal plant cell, therefore water enters by osmosis. The vacuole swells up and presses the cytoplasm against the cell wall, which is made of cellulose fibres laid down in a basket-like arrangement (see Figure 2.16 on page 14). The cell wall stretches slightly and presses back, preventing the cell from bursting. Cells in this swollen condition are said to be **turgid** (see Figure 2.17 on page 14). A young plant depends on the turgor of its cells for support.

Since the dilute sugar solution used in Figure 2.17 has the same water concentration as the cell contents, no net flow of water occurs and the cell remains unchanged (also see Figure 2.18 on page 14).

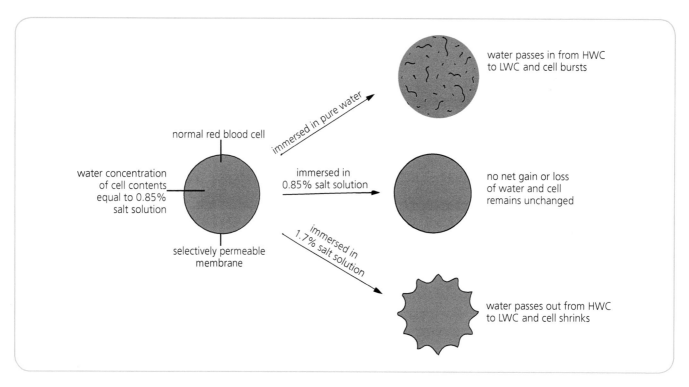

Figure 2.12 Osmosis in a red blood cell (HWC = region of high water concentration, LWC = region of low water concentration)

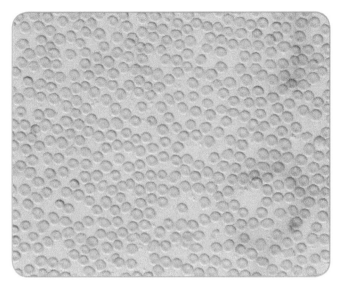

Figure 2.13 Red blood cells in 0.85% salt solution

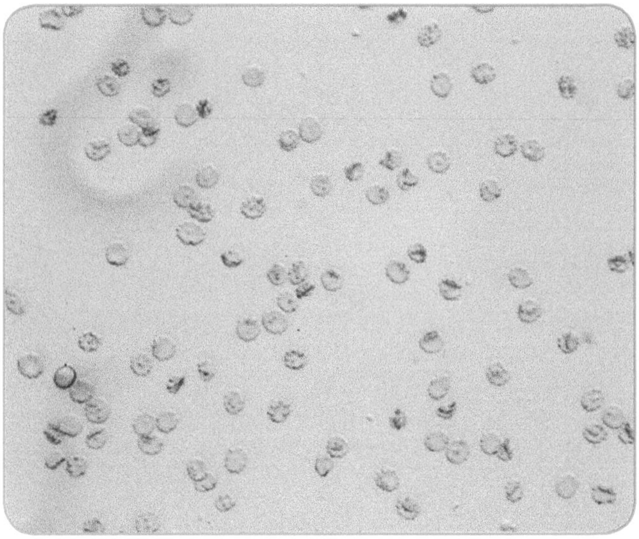

Figure 2.14 Red blood cells in 1.7% salt solution

Related Topic

Control of water balance in unicellular animals

Unicellular animals such as *Amoeba* that live in fresh water take in water continuously by osmosis. Bursting is prevented by the **contractile vacuole** (see Figure 2.15) removing excess water.

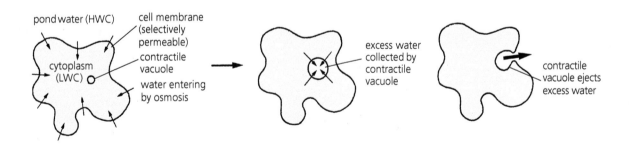

Figure 2.15 Role of contractile vacuole in *Amoeba* (HWC = region of high water concentration, LWC = region of low water concentration)

Since the concentrated sugar solution used in Figure 2.17 has a lower water concentration than the cell contents, water passes out of the cell by osmosis. The living contents shrink and pull away from the fairly rigid cell wall. Cells in this state are said to be **plasmolysed** (also see Figure 2.19). However, they are not dead. When immersed in water, plasmolysed cells regain turgor by taking water in by osmosis.

Figure 2.16 Fibrous nature of cell wall

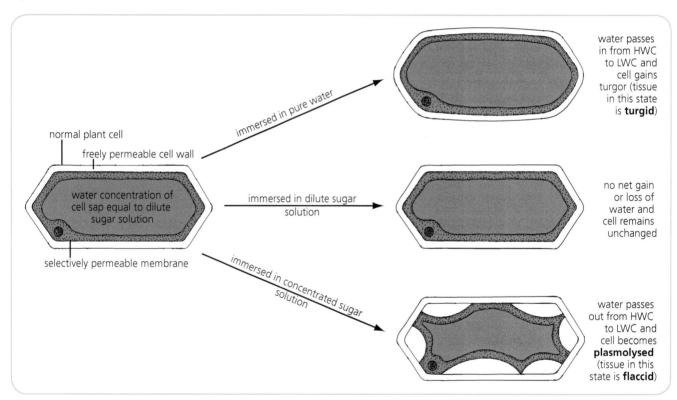

Figure 2.17 Osmosis in a plant cell (HWC = region of high water concentration, LWC = region of low water concentration)

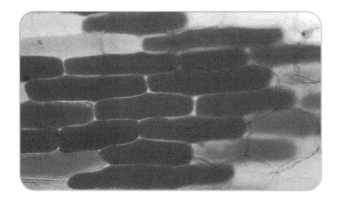

Figure 2.18 Red onion cells unchanged in dilute sugar solution

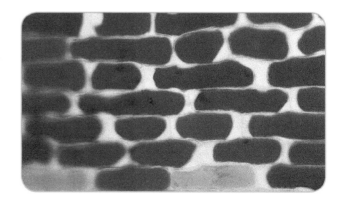

Figure 2.19 Red onion cells plasmolysed in concentrated sugar solution

Power generation by osmosis

Methods of generating power using osmosis are still in the early stages of development. One type of **osmotic power plant** (built in Norway at a location close to where a river meets the sea) is illustrated in a simple way in Figure 2.20. Its successful operation depends on the difference in water concentration between sea water and river water (both of which are readily available).

As liquid rises in the inner compartment and exerts a pressure, this is made to turn a turbine and **generate electrical energy**. The waste product of the process is dilute sea water, which is safely discharged into the surrounding waters.

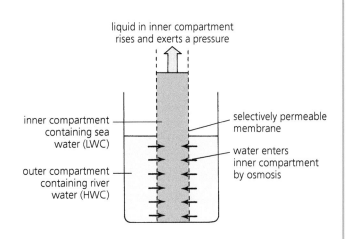

Figure 2.20 Power generation by osmosis

Desalination

Desalination is the process by which the solute (salt) in sea water is separated from its solvent (water) so that the water can be used for drinking or irrigation. It is an important process in countries that have little or no natural fresh water. One method of desalination is reverse osmosis.

Reverse osmosis is the process by which water molecules are forced to pass from sea water, a region of low water concentration (LWC) through a selectively permeable membrane to a region of higher water concentration (HWC). For this to be possible, a pressure has to be applied to the sea water that is far greater than the pressure that would normally be exerted by the water molecules moving passively by osmosis from a HWC to a LWC. Reverse osmosis is shown in a simple way in Figure 2.21.

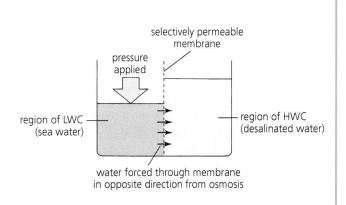

Figure 2.21 Desalination

Active transport

Active transport is the movement of ions (electrically charged particles) or molecules across the cell membrane from a low to a high concentration **against** a concentration gradient. Figure 2.22 on page 16 shows ion type A being actively transported into the cell and ion type B being actively transported out. This form of transport is brought about by certain membrane proteins acting as **carrier molecules** that recognise specific ions or molecules and transfer them across the membrane.

Active transport works in the **opposite direction** to the passive process of diffusion and always requires **energy** for the membrane proteins to move ions or molecules against their concentration gradient.

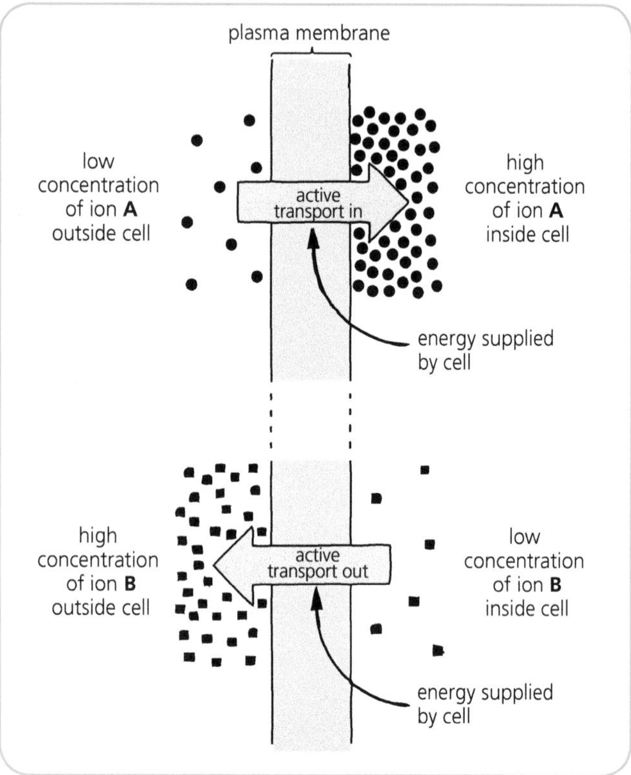

Figure 2.22 Active transport of two different ions

Conditions required for active transport

Active transport requires energy. Therefore factors such as temperature and availability of oxygen and food, which directly affect a cell's respiratory rate, also affect the rate of active transport.

Iodine

Sea water contains only a tiny trace of **iodine**, yet some brown seaweeds are found to have 0.4% of their dry weight as iodine. By actively transporting iodine into their cells against a concentration gradient, these seaweeds are able to concentrate iodine in their cell sap by a factor of **several thousand** times. Therefore they provide a commercial source of this element.

Uptake of dyes by yeast cells

A small sample of yeast cells from a freshly aerated culture is placed on a slide and a drop of **methylene blue dye** is added. The procedure is repeated using a sample of yeast cells that have been boiled and cooled. After 5 minutes, cover slips are added and the cells are viewed under the microscope.

The live cells are found to contain little or no blue dye but the dead cells are found to be blue. Therefore it can be concluded that the blue dye diffuses into the yeast cells but only the live cells (that are able to respire and generate energy) are able to actively transport it back out against a concentration gradient.

Investigation

Testing Your Knowledge 2

1 Rewrite the following paragraph choosing the correct answer at each underlined choice. (8)

Osmosis is a form of diffusion/ion uptake. It is an active/a passive process that requires/does not require energy. During osmosis oxygen/water molecules move through a freely/selectively permeable membrane from a higher/lower water concentration to a higher/lower water concentration down/against a concentration gradient.

2 a) With reference to the water concentrations involved, explain why a red blood cell bursts when placed in water yet an onion epidermal cell does not. (4)

b) Why does a red blood cell shrink when placed in concentrated salt solution? (1)

3 a) Describe and explain the osmotic effect of very concentrated sugar solution on an onion epidermal cell. (2)

b) What term is used to describe a cell in this state? (1)

c) How could such a cell be restored to its turgid condition? (1)

4 a) Define the term *active transport*. (2)

b) Copy and complete Table 2.1 by answering the five questions. (5)

Question	Diffusion	Active transport
What is an example of a substance that moves through the cell membrane by this process?		
Do the particles move from high to low or from low to high concentration?		
Do the particles move down or against a concentration gradient?		
Is energy required?		
Is the process passive?		

Table 2.1

What You Should Know Chapters 1–2

active	gain	phospholipid
against	gradient	plasmolysed
animal	lose	ribosomes
burst	lower	sap
cells	membrane	selectively
chloroplasts	mitochondria	shrink
concentration	nucleus	supported
diffusion	osmosis	useful
down	passive	vacuole
energy	permeable	walls

1 All living things are composed of one or more _____, the basic units of life.

2 The cells of green plants, animals and fungi have several structures in common including a _____, a cell _____ and mitochondria.

3 _____ cells lack a cell wall and a large central _____.

4 Only green plant cells contain _____ but all cells contain ribosomes.

5 The cell's activities are controlled by the nucleus. A plant cell is _____ by the cell wall made of cellulose fibres. The _____ is stored in the central vacuole.

6 Chloroplasts are responsible for photosynthesis, _____ for respiration and _____ for protein synthesis.

7 The cell membrane consists of protein and _____ molecules and is _____ permeable.

8 Diffusion is the movement of molecules from a high to a low concentration _____ a concentration _____. It is a _____ process and does not require a supply of energy.

9 _____ is the means by which _____ molecules enter and waste materials leave a cell.

10 _____ is a special case of diffusion during which water molecules move along a water _____ gradient from a region of higher water concentration to a region of _____ water concentration through a selectively _____ membrane.

11 When placed in a solution with a water concentration higher than that of their cell contents, cells _____ water by osmosis. Animal cells swell up and may _____; plant cells also swell up but are prevented from bursting by their cell _____.

12 When placed in a solution with a water concentration lower than that of their cell contents, cells _____ water by osmosis. Animal cells _____ and plant cells become _____.

13 Ions and molecules of some chemical substances are actively transported across the cell membrane _____ a concentration gradient.

14 _____ transport is brought about by membrane protein molecules and requires _____.

DNA and the production of proteins

Structure of DNA

Chromosomes are threadlike structures composed of genes found inside the nucleus of plant and animal cells. They contain **DNA** (deoxyribonucleic acid). A molecule of DNA consists of two strands twisted together into a **double helix**. Each strand has a 'chain' of **base** molecules called adenine (A), thymine (T), guanine (G) and cytosine (C) (see Figure 3.1).

Normally the two strands are held together by weak bonds between pairs of the bases. Each base's molecular structure is such that it can only fit with one other type of base. A fits with T and G fits with C. **A-T** and **G-C** are called **complementary base pairs**.

Sequence of DNA bases

The DNA of one member of a species differs from that of another member by the order in which the bases occur in their chromosomes. It is this **sequence** of bases along the DNA strands that is unique to an organism. The sequence contains the **genetic instructions** that control the organism's inherited characteristics.

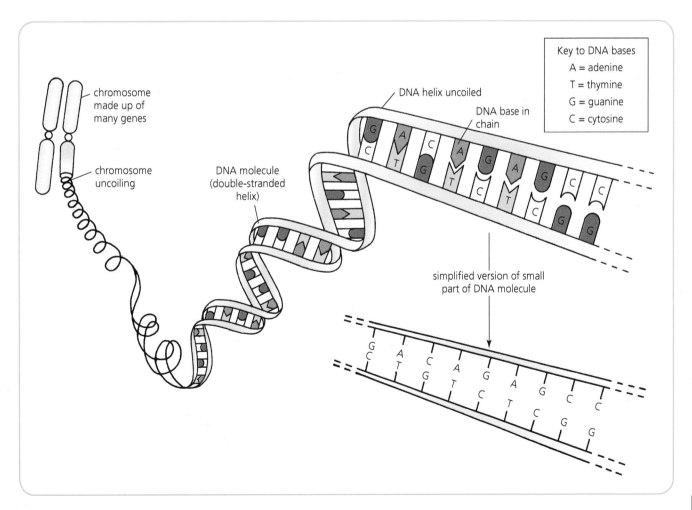

Figure 3.1 Structure of DNA

Genetic code

A section of DNA on a chromosome that codes for a protein is called a **gene** and normally it is hundreds or even thousands of bases long. The information present in the DNA takes the form of a molecular language called the **genetic code**. Amino acid molecules are the building blocks of proteins. Each group of three bases along a DNA strand represents a **codeword** for one of the 20 types of amino acid that make up proteins.

| **Research Topic** | **Establishing the structure of DNA** |

Chemical analysis

In the late 1940s, Chargaff analysed the base composition of DNA extracted from a number of different species. He found that the quantities of the four bases were not all equal but that they always occurred in a **characteristic ratio** regardless of the source of the DNA.

These findings, called **Chargaff's rules**, are summarised as follows:

- The number of A bases = the number of T bases (in other words A:T = 1:1).
- The number of G bases = the number of C bases (in other words G:C = 1:1).

X-ray crystallography

At around the same time, Wilkins and Franklin were using **X-ray crystallography** to investigate the structure of DNA. They found that when X-rays are passed through a crystal of DNA, they become deflected into a **scatter pattern** (see Figure 3.2). The form this pattern takes is determined by the arrangement of the atoms in the DNA molecule.

When the scatter pattern is recorded using a photographic plate (see Figure 3.3), it reveals information that can be used to build up a **three-dimensional picture** of the molecules in the crystal. Wilkins and Franklin found that the X-ray scatter patterns of DNA from different species were identical.

From the X-ray scatter patterns produced by Wilkins and Franklin, Watson and Crick figured out that the DNA must be a long, thin molecule of constant diameter coiled in the form of a **helix**. From Chargaff's rules they figured out that base A must be paired with base T and base G with base C. They realised that this could only be possible if DNA consisted of two strands held together by specific **pairing of bases**. Watson and Crick then set about building a wire model of DNA and in 1953 were first to establish the three-dimensional **double helix** structure of DNA.

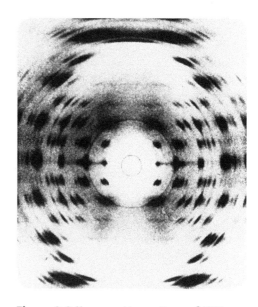

Figure 3.3 X-ray scatter pattern of DNA

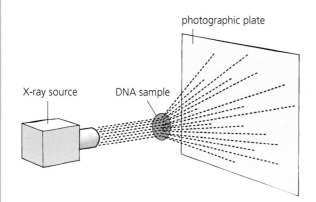

Figure 3.2 X-ray crystallography

Messenger RNA (mRNA)

A gene codes for a particular protein by making a messenger molecule that is complementary to one of its DNA strands (see Figure 3.4). This molecule is called **messenger RNA (mRNA)**. It carries the copy of the genetic code from the gene's DNA in the nucleus out into the cytoplasm. When mRNA meets a **ribosome**, amino acid molecules become assembled into a molecule of protein.

The sequence of amino acids that are joined together into the protein is determined by the information in the sequence of the mRNA's codewords and (indirectly) by the order of the bases on the DNA. By this means, DNA encodes the information for a particular **sequence** of amino acids in a protein, which in turn dictates the **structure** and **function** of that protein (see Chapter 4).

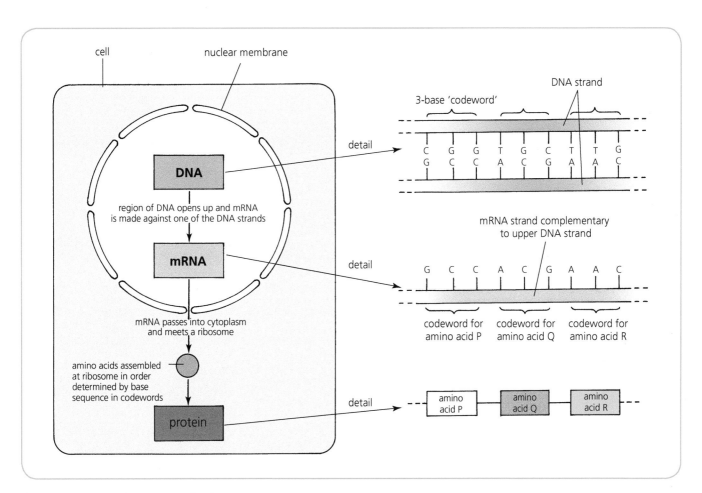

Figure 3.4 DNA and protein production

Figure 3.5 shows the diploid chromosome complement of a fruit fly. These chromosomes (extracted from cells of larvae) are found to take up stain and produce a **banding pattern** that is a **constant characteristic** of each type of chromosome. This banding pattern is shown for chromosome 1.

It is possible to relate the **location of individual genes** to particular bands on chromosomes. For example, when chromosome 1 from white-eyed flies is compared with chromosome 1 from red-eyed flies (see Figure 3.6), the former is found to lack one of the characteristic bands. Since the chromosome complement of the two types of fly is identical in every other way, it is concluded that the gene that controls red eye colour must be located at the position of the missing band on chromosome 1.

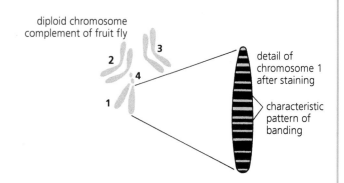

Figure 3.5 Pattern of banding on chromosome

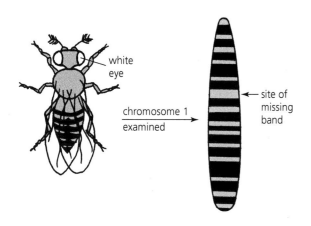

Figure 3.6 Location of gene on chromosome

Testing Your Knowledge

1 What are chromosomes and where are they located in a cell? (2)

2 a) What do the letters DNA stand for? (1)
  b) How many strands are present in a DNA molecule? (1)
  c) i) How many different types of base molecule are found in DNA?
    ii) Describe the base-pair rule. (2)

3 a) By what means is the genetic code present in the DNA of a gene transported from the nucleus to the cytoplasm of a cell? (1)

b) i) What subunits make up a protein?
   ii) Where in the cytoplasm are these subunits assembled into proteins?
   iii) What determines the sequence in which these subunits are joined together?
   iv) Why is the sequence of these subunits in a protein molecule important? (4)

4 Proteins

Amino acids

Each **protein** is built up from a large number of subunits called **amino acids** of which there are **20** different types. These are joined together into chains. Each chain normally consists of hundreds of amino acid molecules linked together. Depending on which amino acids are in the chain, further bonds between certain amino acids form. These make the chain coil up and become folded in a **characteristic way**. The exact nature of the coiling and folding depends on the **sequence** of the amino acids present in the chain. It determines the final **structure** of the protein, which in turn determines the function it will carry out.

Molecular shape

The **shape** of protein molecules shows wide variation. For example, an enzyme molecule has a roughly spherical shape (see Figure 4.1) whereas the protein molecules that make up structures such as bone and connective tissues often take the form of rope-like fibres. The proteins that form part of the cell membrane (see Figure 2.3) tend to be spherical or ovoid and those that make up antibodies are Y-shaped (see Figure 4.2).

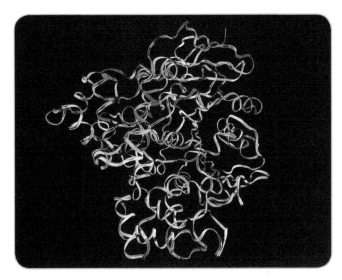

Figure 4.1 Computer-generated model of an enzyme molecule

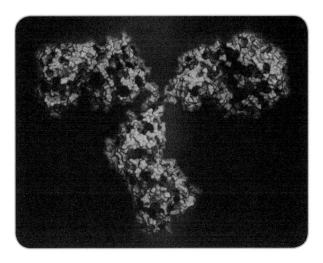

Figure 4.2 Antibody molecule

Variety of functions

Proteins as a group perform a wider range of **functions** than any other type of molecule present in the human body. Some are found in bone and muscle where their strong fibres provide support or allow movement. Others play a variety of essential roles as follows.

Enzymes

Figure 4.3 (overleaf) shows an **enzyme** molecule. It is made up of several chains of amino acids folded and coiled in a particular way that exposes an **active site**. This site is able to combine readily with a substrate (see page 26) and speed up a biochemical reaction such as respiration.

Structural proteins

Protein is an essential **structural component** of the cell membrane and of the membrane that encloses each of the cell's organelles.

Hormones

Hormones are chemical messengers transported in an animal's blood to 'target' tissues where they bring about a particular effect. Many hormones are made of protein and **regulate** processes such as growth and metabolism.

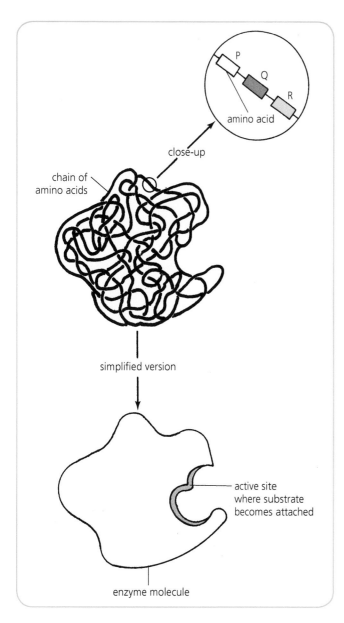

Figure 4.3 An enzyme's molecular structure

Antibodies

Antibodies are protein molecules made by white blood cells to **defend** the body against micro-organisms that could cause disease.

Receptors

Cells communicate with one another using **signal molecules**. A cell that is the target of a signal molecule has a type of **receptor protein** (normally in its membrane) that recognises and is a perfect fit for the signal molecule. The signal becomes attached to the receptor which responds by triggering a chemical event. For example, protein receptors present in the membrane of a nerve cell detect chemical signals released by other nerve cells and enable nerve impulses to be transmitted on through the nervous system.

Biological catalysts

Since living things cannot tolerate the high temperatures needed to make chemical reactions proceed at a rapid rate, they employ catalysts. **Biological catalysts** are called **enzymes**. Enzymes are made by all living cells.

Importance of enzymes

Enzymes **speed up** the rate of all biochemical reactions yet remain **unchanged** by the process. They allow biochemical reactions to proceed rapidly at the relatively low temperatures (such as 5–40 °C) needed by living cells to function properly. Some are involved in degradation reactions (the breaking down of complex molecules); others control synthesis reactions (the building-up of complex molecules). In the absence of enzymes, biochemical pathways such as respiration and photosynthesis would proceed so slowly that life as we know it could not exist.

Importance of a control

A **control** is a copy of the experiment in which all factors are kept exactly the same except for the one being investigated in the original experiment. When the results are compared, any difference found between the two must be due to that **one factor**. For example, in the experiments shown on page 25, we can conclude that in the first one a catalyst and in the second one an enzyme increased the rate of breakdown of hydrogen peroxide. If controls had not been set up, it would be valid to suggest that the breakdown of hydrogen peroxide would have proceeded rapidly whether a catalyst or an enzyme had been present or not.

Effect of a catalyst on the breakdown of hydrogen peroxide

In the experiment shown in Figure 4.4, the bubbles forming the froth in tube A are found to relight a glowing splint. This shows that oxygen is being released during the breakdown of hydrogen peroxide. In tube B, the control, the breakdown process is so slow that no oxygen can be detected.

It is concluded therefore that manganese dioxide (which remains chemically unaltered at the end of the reaction) has increased the rate of this chemical reaction, which would otherwise have only proceeded very slowly. A substance that has this effect on a chemical reaction is called a **catalyst**.

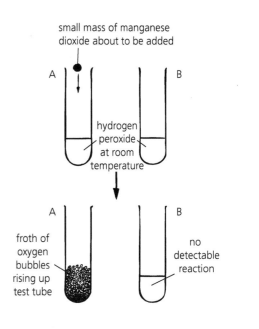

Figure 4.4 Effect of a catalyst

Effect of catalase on the breakdown of hydrogen peroxide

Catalase is an enzyme made by living cells. It is especially abundant in fresh liver cells. In the experiment shown in Figure 4.5, the bubbles produced in tube C are found to relight a glowing splint. This shows that oxygen is being released during the breakdown of hydrogen peroxide. In tube D (the control), boiling the liver has destroyed the catalase with the result that the breakdown process is too slow for any oxygen to be detected.

It is concluded that the enzyme catalase has increased the rate of the chemical reaction shown in the following equation, which would otherwise have proceeded only very slowly.

hydrogen peroxide $\xrightarrow{\text{catalase}}$ oxygen + water
   (substrate)   (end products)

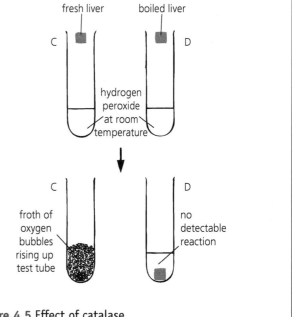

Figure 4.5 Effect of catalase

Active site

At some point on the surface of the protein molecule that makes up an enzyme, there is an **active site** (see Figure 4.6). The shape of the active site is determined by its chemical structure, which results from the **sequence** of, and **bonding** between, **amino acids** in the enzyme molecule.

Mechanism of action

An enzyme is able to act on only one type of substance (its **substrate**) since this is the only substance whose molecules exactly fit the enzyme's active site. The enzyme is therefore said to be **specific** to its substrate and the shape of the enzyme's active site is said to be **complementary** to the substrate's molecular shape.

When a molecule of substrate enters the active site, the shape of the enzyme molecule and its active site change slightly, making the active site **fit closely** round the substrate molecule. This close contact increases the chance of the chemical reaction taking place (see Figure 4.6).

Factors affecting enzyme activity

To function efficiently an enzyme requires a suitable **temperature** and a suitable **pH**. Each enzyme is found to be most active in its optimum (best) conditions.

One variable factor

An investigation is a fair test if at each stage only one difference (**variable factor**) is studied at a time. If several differences were involved at the same time then it would be impossible to know which one was responsible for the results obtained. The experiment shown in the Related Activity opposite is **valid** and **fair** because it tests only one variable factor (temperature). (Also see Appendix 4.)

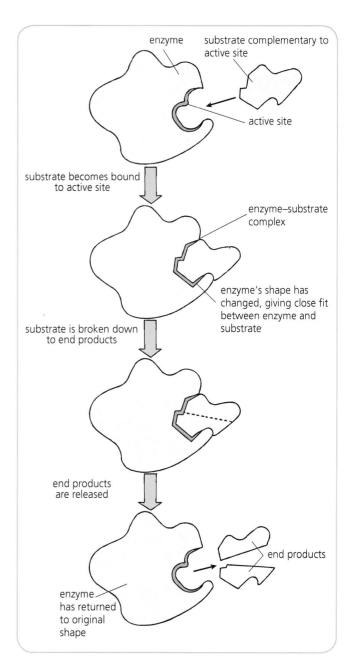

Figure 4.6 Enzyme-catalysed reaction

Effect of temperature

The graph in Figure 4.8 on the next page summarises the general effect of **temperature** on enzyme activity. At very low temperatures, enzyme molecules are **inactive** but **undamaged**. At low temperatures, enzyme and substrate molecules move around **slowly** in their surrounding medium. They meet only rarely and the rate of enzyme activity remains low. As the temperature increases, the two types of molecule move about at a **faster rate** and the rate of the reaction increases.

Optimum temperature

The temperature at which **most enzyme-substrate molecular activity** occurs (without any molecular damage taking place) is within the range of 35–40 °C. This temperature at which the reaction works best is called the enzyme's **optimum** temperature.

Testing Your Knowledge 1

1 Copy the following sentences and complete the blanks. The amino acids in a protein are built into a particular order that is determined by the sequence of the _____ on a portion of DNA in a _____. The sequence of amino acids determines the protein's _____ and _____. (4)

2 a) What general name is given to biological catalysts? (1)
  b) State TWO properties of a biological catalyst. (2)

3 a) Where are enzymes found in a living organism? (1)
  b) Of what type of substance are enzyme molecules composed? (1)

  c) Briefly explain why enzymes are needed for the functioning of all living cells. (2)

4 a) What determines the shape of an enzyme's active site? (1)
  b) What term is used to refer to the substance upon which an enzyme acts? (1)
  c) Why is an enzyme said to be *specific* in its relationship with its substrate? (1)

Related Activity

Investigating the effect of temperature on amylase activity

Amylase is an enzyme that promotes the breakdown of **starch** (its substrate) to simple sugar. In the experiment shown in Figure 4.7 overleaf, a test tube containing amylase and a test tube containing starch suspension are placed in each of four water baths at **different temperatures**. The test tubes are left for 30 minutes to allow their contents to equilibrate (become equal with) the temperature of the surrounding water.

During this time, two spotting tiles are set up with a drop of **iodine solution** in every cavity. Following equilibration, the enzyme in each tube is added to the tube of starch at each

temperature. Separate droppers are immediately used to add a small sample of each tube's **enzyme-substrate mixture** to iodine on the spotting tile. This procedure is repeated at the time intervals indicated in Figure 4.7, which also shows a typical set of results after 20 minutes.

Failure by a sample to produce a blue-black colour with iodine solution is regarded as evidence that the starch in that tube has been broken down by amylase. Since failure to produce the blue-black colour occurs first with the samples from the tube at 35 °C, it is concluded that the amylase is most active at 35 °C. At the higher temperature of 60 °C, no breakdown of starch is found to occur. At the lower temperature of 2 °C, breakdown of starch takes a long time.

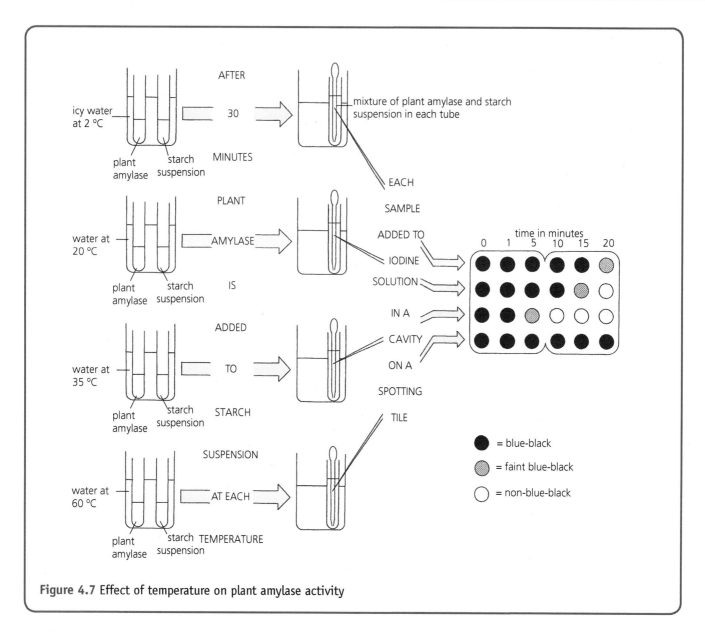

Figure 4.7 Effect of temperature on plant amylase activity

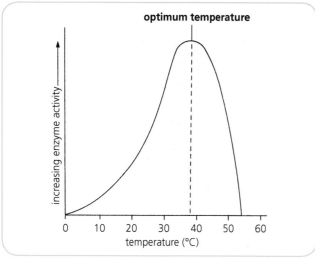

Figure 4.8 Effect of temperature on enzyme activity

Denaturation

At temperatures above 40 °C, an enzyme's atoms vibrate so much that some of the chemical bonds holding its amino acids together break and the molecule begins to come apart. Soon the shape of its active site becomes **altered** and the enzyme is unable to fit with its substrate. An enzyme in this damaged state is said to be **denatured**. It is permanently inactive.

Beyond 40 °C, as more and more molecules of enzyme become denatured and inactive, the rate of the reaction **decreases rapidly**. At temperatures of about 55–60 °C enzyme activity is often found to have come to a complete halt. This is because all the enzyme molecules have become denatured.

Related Activity

Investigating the effect of pH on catalase activity

In the experiment shown in Figure 4.9, the action of **catalase** on hydrogen peroxide is investigated. The one variable factor is the **pH** of the hydrogen peroxide solution. Each of the different pH conditions is maintained by adding a suitable **buffer solution** (a special chemical that keeps a solution at a required pH).

When an equal-sized piece of fresh liver was added to each cylinder, the results shown in the diagram were produced. This is because liver contains the enzyme catalase, which promotes the breakdown of hydrogen peroxide to oxygen and water. As oxygen is released, it produces a **froth of bubbles**. The height of the froth formed indicates the activity of the enzyme at each pH.

From the results it can be seen that this catalase was most active at around pH 9 and worked fairly well over a range of about pH 7 to 11. This is called its **working range** of pH. This catalase does not work well outside this range.

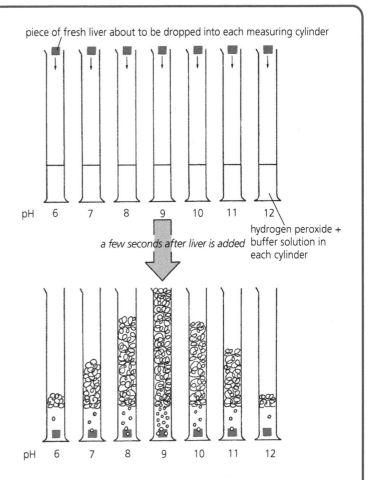

Figure 4.9 Effect of pH on catalase activity

Related Activity

Investigating the effect of pH on the activity of pepsin

Pepsin is an enzyme secreted by glands in the stomach wall. It digests **protein** to soluble end products. The protein used in the experiment shown in Figure 4.10 is egg white. It has been heated in advance to form a cloudy suspension of insoluble particles ready for use. The three test tubes are set up as shown in the diagram and maintained at 37 °C, the optimum temperature for pepsin activity.

After 30 minutes, the contents of tube A (acidic conditions) are found to be **clear** whereas those of B (neutral conditions) and C (alkaline conditions) remain **cloudy** and unchanged. It is therefore concluded that pepsin digests large particles of insoluble protein to soluble end products in **acidic** conditions only.

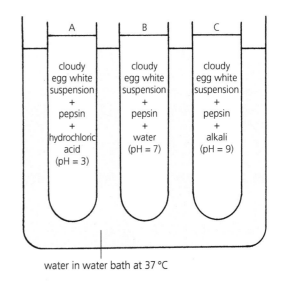

Figure 4.10 Effect of pH on pepsin

Optimum pH

Each enzyme is found to be most active at a particular pH (its **optimum** pH), as shown in Figure 4.11. Most enzymes function within a working pH range of about 5–9 with an optimum of around pH 7 (neutral). However, there are exceptions. Pepsin works best in strongly acidic conditions of pH 2.5; alkaline phosphatase (which plays a role in bone formation) works best at pH 10.

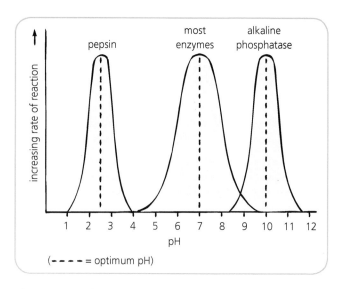

Figure 4.11 Effect of pH on enzymes

Related Activity

Effect of temperature and pH on egg white

Egg white is made of a protein called albumin. In this activity, the albumin acts as a **model** for the effect of temperature and pH on proteins in general, including **enzymes**.

Temperature

When raw egg white is heated, the albumin coagulates and changes from a colourless, transparent liquid to a white, opaque solid. On cooling, the egg white protein does not return to its original state. Its molecular structure has been altered irreversibly (**denatured**).

pH

When acid is added to raw egg white to reduce its pH value to a low level, the albumen coagulates. When the pH is returned to neutral by the addition of an alkali, the egg white protein does not resume its original transparent state. The change in its molecular structure is irreversible and it is in a denatured state.

Similarly, most enzymes are irreversibly altered in molecular structure by high temperatures and extremes of pH and therefore become non-functional following exposure to these adverse conditions.

Testing Your Knowledge 2

1 Salivary amylase is an enzyme that digests starch in the human mouth.
   a) Compare the rate of amylase activity at 5 °C and 25 °C. (1)
   b) i) Compare the rate of amylase activity at 40 °C and 70 °C.
      ii) Explain the difference in rate in terms of the molecules. (3)

2 a) Which temperature given in question **1** is closest to the optimum for most enzymes? (1)
   b) i) Suggest a temperature at which an enzyme is inactive but capable of activity if the temperature changes.
      ii) Explain your answer. (2)

3 a) Explain the meaning of the term *optimum condition* as applied to the activity of an enzyme. (1)
   b) Which of the following is the optimum range of pH for the human enzyme pepsin? (1)
      A 2–3
      B 4–5
      C 6–7
      D 8–9

5 Genetic engineering

Chromosomal material of a bacterium

A bacterium has one **chromosome** in the form of a complete circle and one or more small circular **plasmids,** as shown in Figure 5.1. The chromosome and plasmids are composed of DNA and are made up of many genes. Each gene carries the information necessary for the production of a certain protein such as an enzyme.

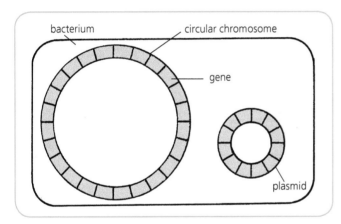

Figure 5.1 Genetic material of a bacterium

Genetic engineering

Genetic information in the form of DNA can be transferred from one cell to another through a process called **genetic engineering**. In recent years, scientists have developed technology that enables them to transfer pieces of chromosome containing sections of DNA from one organism such as a human being to another organism such as a bacterium. An organism that has been 'reprogrammed' is described as a **genetically modified (GM)** organism. This transformed organism now possesses some characteristic that makes it useful to humans.

For example, a GM bacterium may act as a chemical factory and produce a useful product, as shown in Figure 5.2 (oveleaf). This diagram shows a simplified version of genetic engineering involving the following stages:

1 identification of the section of DNA that contains the required gene (such as the gene for human insulin) on the source chromosome
2 cutting of the source chromosome using a special enzyme (acting as biochemical 'scissors') to extract the required gene
3 extraction of a plasmid from a bacterial cell
4 opening up of a plasmid using the same enzyme used in stage 2
5 insertion and sealing of the required gene into the bacterial plasmid using a different type of enzyme
6 insertion of the plasmid into the bacterial host cell
7 growth and multiplication of the genetically modified bacterial cell
8 formation of many duplicate plasmids that express the 'foreign' gene and make the desired useful product (such as insulin), which is then extracted and purified.

In this example the plasmid containing the required gene acts as a carrier between the two different organisms.

Advantage of using micro-organisms

Whereas isolated cells of an advanced multicellular organism such as a human being are often difficult to mass-produce in culture, micro-organisms can be grown quickly and easily, often at low cost. Given suitable conditions, GM bacteria multiply at a rapid rate and manufacture large quantities of a **useful product** that can then be extracted, concentrated, purified and put to use.

Multicellular organisms

The production of GM organisms by genetic engineering is not restricted to micro-organisms such as bacteria and yeast. In recent years many applications

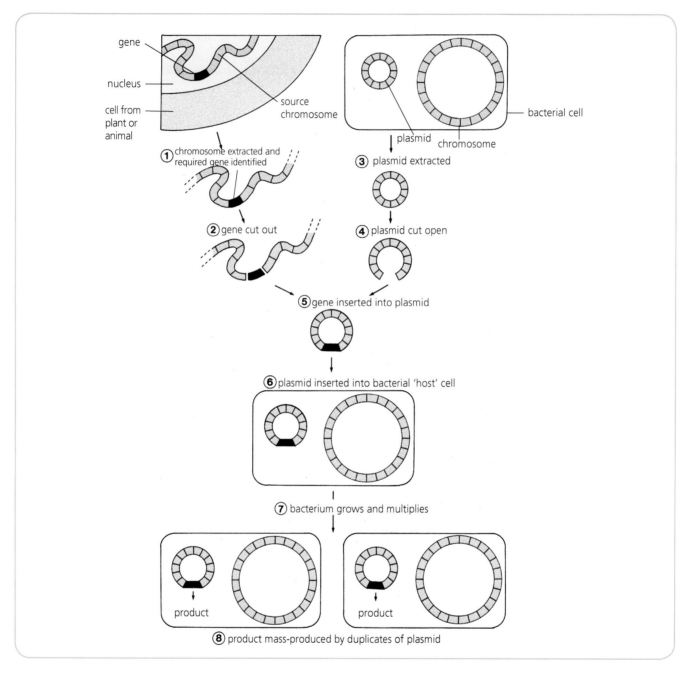

Figure 5.2 Genetic engineering

have been successfully carried out involving multicellular organisms. For example, scientists have found ways of injecting a plasmid containing a required gene into the cells of a multicellular organism. These transformed cells then grow into a **multicellular GM organism**. By this means, genetic engineers have produced:

- plants with improved nutritional value, such as golden rice (see Related Topic opposite)
- plants that are resistant to pests, such as blight-resistant potatoes (see Related Topic on page 34)
- fruit with a longer shelf life, such as tomatoes (see Related Topic on page 34).

Products of medical value

Many products of genetic engineering have important **medical applications**. In each of the examples given in Table 5.1, a particular gene has successfully been inserted into a bacterium, which is then allowed to multiply and make the product of medical value.

Scientists have also managed to reprogramme yeast cells to produce **human serum albumin** (used in blood replacements), **epidermal growth factor** (which speeds

up the healing of wounds) and **hepatitis B antigens** (used in the manufacture of hepatitis B vaccine).

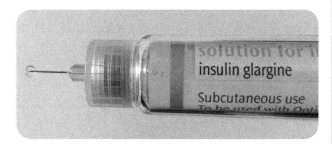

Figure 5.3 Insulin pen

Product of genetic engineering	Normal source and function of substance	Medical application of gene product made by GM bacteria
insulin	Made by pancreas cells. Controls the level of glucose in the blood.	Given to people who are diabetic and do not make enough insulin naturally. (An insulin pen is shown in Figure 5.3.)
human growth hormone (HGH)	Made by cells in the pituitary gland. Essential during childhood and adolescence to control growth and development.	Given by regular injection to children who do not make enough of their own. Prevents severely reduced growth.

Table 5.1 Medical applications of gene products

Golden rice

Vitamin A is a chemical needed in small quantities in the diet for good health and, in particular, for healthy eyesight. Normal cultivated rice contains very little vitamin A in its grains. In regions of the world where this type of rice is the main part of the staple diet and where people fail to obtain enough vitamin A from other food sources, vitamin A deficiency is common. Shortage of this vitamin leads to **vision problems** and even blindness and may cause premature death.

Genetic engineers have now succeeded in transferring two genes from other plants into traditional rice to create a new variety called **golden rice** (see Figure 5.4). This new strain of rice plant makes a yellow chemical that accumulates in the rice grains giving them a golden appearance. When the rice grains are eaten, the yellow chemical is changed into vitamin A by the body. It then acts in the normal way and prevents vision problems.

Golden rice's nutritional value is further increased by the presence of a higher concentration of **iron** in its grains compared with those of normal varieties. Iron is needed for the production of new red blood cells.

Figure 5.4 Golden rice

Blight-resistant potatoes

A fungal parasite attacks potato plants and causes a disease called **blight** (see Figure 5.5) that ruins the potatoes and makes them inedible. It was this devastating disease that led to the Irish potato famine in 1845–6 and the Scottish Highland potato famine in 1846.

In recent years, genetic engineers managed to identify and extract from a wild potato plant a gene that gave **blight resistance**. They then introduced it into other varieties of potato plant, making them resistant. However, new strains of the fungus soon evolved that were able to break down the potato plants' resistance.

Disease resistance is now known to be a complex feature of a potato plant's genetic makeup. It involves **several interacting genes**. Some Scottish scientists are using a technique that enables them to locate several genes (some for disease resistance, others for quality traits) and transfer them as a group. On being introduced to potato plants, these act together and significantly increase the plant's resistance to blight.

The production of GM crops that are resistant to pests is of benefit to the environment because little or no chemical pesticide needs to be used to keep the crops healthy.

Figure 5.5 Potato leaves infected with blight

Tomatoes with longer shelf life

Genetic engineers are now able to insert into an organism a gene that **switches off** (silences) the activity of another gene. This process has been developed commercially to extend the **shelf life** of tomatoes. Under normal circumstances, a certain gene controls the production of an enzyme that breaks down a chemical in the cell walls of a tomato, making the fruit become soft.

By switching off the gene that codes for this enzyme (but not the genes that control colour or flavour), scientists have made it possible to **delay the softening** of tomatoes while allowing their colour and flavour to develop as normal. Therefore the fruit has a much longer shelf life before it becomes over-ripe and spoiled. The two groups of tomatoes shown in Figure 5.6 are the same age and are genetically identical except that those on the top have been genetically modified to silence the 'fruit-softening' gene.

Figure 5.6 GM tomatoes (top) and their non-GM relatives

Testing Your Knowledge

1 a) What is meant by the term *genetic engineering*? (2)
   b) An agent that acts as a carrier between two species is often called a vector. Which part of a bacterium's genetic material could this term be used to describe? (1)

2 Arrange the following steps involved in the process of genetic engineering into the correct order. (1)
   A cutting of source chromosome to extract required gene
   B insertion of required gene into plasmid and plasmid into host cell
   C growth of transformed host cell into GM organism
   D extraction of a plasmid from a bacterial cell
   E identification of section of DNA that contains required gene

3 a) Give an example of a product of genetic engineering. (1)
   b) Explain why this product is of benefit to human beings. (1)

What You Should Know Chapters 3–5

acids	DNA	plasmid
active	engineering	products
amino	enzymes	ribosome
antibodies	function	RNA
base	gene	sequence
benefit	genetic	shape
catalyst	helix	site
chromosome	host	structural
code	inserted	substrate
complementary	modified	temperature
denatured	optimum	unaltered

1 Chromosomes contain DNA. A molecule of DNA consists of two strands coiled into a double-stranded _____. The two strands are held together by weak bonds between their bases. There are four types of base. A pairs with T and G pairs with C to form complementary _____ pairs.

2 Information called the genetic _____ is present in DNA. It takes the form of codewords of bases that correspond to _____ acids, the building blocks of proteins.

3 The sequence of bases in DNA determines the sequence in which amino _____ are assembled in a protein.

4 The genetic code for a protein, present on DNA, is carried by messenger _____ from the nucleus to a _____ in the cytoplasm where the appropriate amino acid become joined together into the protein.

5 Both the shape and _____ of a protein molecule are determined by the particular _____ of amino acids that make up the protein.

6 Some proteins have a _____ function, others act as hormones, receptors, _____ or enzymes.

7 A _____ is a substance that speeds up the rate of a chemical reaction but remains _____ by the reaction. _____ are biological catalysts produced by all living cells.

8 The shape of the _____ site on an enzyme molecule is _____ to the molecular structure of its _____ allowing them to combine together closely.

9 Following catalytic activity, the end _____ become detached from the active _____ leaving the enzyme unchanged.

10 To function efficiently, an enzyme needs an appropriate pH and a suitable _____. Each enzyme works best in its _____ conditions. At temperatures above 55 °C, most enzyme molecules are inactive because their molecular _____ has been irreversibly altered. They are described as being _____.

11 _____ can be transferred naturally from one cell to another. DNA can also be transferred by genetic _____.

12 During _____ engineering, the section of DNA that contains the required _____ is identified and cut out of the source _____. Then the gene is inserted into a bacterial _____.

13 The reprogrammed plasmid is _____ into a _____ cell such as a bacterium.

14 The transformed host cell (now a genetically _____ organism) grows and multiplies and provides humans with some _____.

6 Respiration

Glucose

Glucose, a type of sugar, is an energy-rich substance. It is the most common end product from the digestion of complex carbohydrates such as starch. Glucose is the main source of **energy** in a living cell.

Release of chemical energy

When a food is burned in air or in oxygen, its **chemical energy** is released rapidly as heat and light (see the Related Activity below). However, in a living cell, the chemical energy stored in food is not released as rapidly. Instead an orderly release of energy takes place relatively slowly through a series of enzyme-controlled reactions called **respiration**. Respiration occurs in all living cells.

Role of ATP

ATP (adenosine triphosphate) is an energy-rich compound. When glucose is broken down during respiration in a living cell, the energy released is used to generate ATP. As a result, many molecules of ATP are present in every living cell. When required to do so, these ATP molecules immediately transfer their energy and make it available for energy-requiring cellular activities such as:

- muscular contraction
- cell division
- synthesis of proteins
- transmission of nerve impulses
- active uptake of ions and molecules (see pages 15–16)
- carbon fixation during photosynthesis (see page 143).

Related Activity

Measuring the energy content of food

Energy is measured in **kilojoules (kJ)**. The quantity of energy required to raise the temperature of 1000 g of water by 1 °C is 4.2 kJ. The chemical energy contained in food can be changed into heat energy and measured. In the experiment shown in Figure 6.1, 1 g of peanut is ignited and held under a beaker containing 1000 g water. The temperature of the water is found to rise by 2 °C. Therefore 1 g of peanut has released 8.4 kJ of energy.

Sources of error
The experiment in Figure 6.1 contains the following sources of error:

- much of the heat energy released by the burning peanut is lost to the surroundings
- the heat energy that does reach the water is not evenly spread out
- the peanut is not completely burned to ashes.

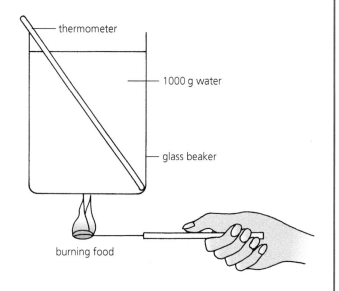

Figure 6.1 Measuring energy in food

Food calorimeter

The energy content of a food can be measured accurately using a **food calorimeter** (see Figure 6.2). This equipment overcomes the limitations of the simple apparatus shown in Figure 6.1 in the following ways:

- the food sample is **enclosed**, therefore loss of heat energy is reduced to a minimum
- the **stirrer** and **coiled chimney** bring about even distribution of the heat energy released during burning
- the **oxygen** supply ensures that the peanut burns completely.

This time the temperature of the water is found to rise by 6 °C, showing that 1 g of peanut has released 25.2 kJ of energy.

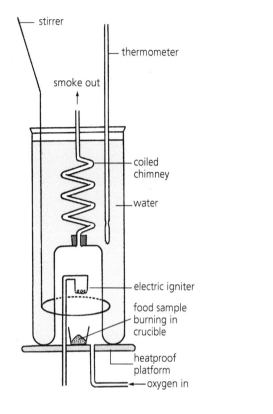

Figure 6.2 Food calorimeter

Related Activity

Investigating respiration in a green plant

Hydrogen carbonate indicator solution is a chemical that varies in colour depending on the concentration of carbon dioxide (CO_2) that it contains. Table 6.1 shows the relative CO_2 concentrations indicated by the various colours.

Relative CO_2 concentration	Colour of hydrogen carbonate indicator
high (above atmospheric)	yellow
medium (atmospheric)	red
low (below atmospheric)	purple

Table 6.1 Hydrogen carbonate indicator range

In the experiment shown in Figure 6.3 overleaf, the hydrogen carbonate indicator in tube A changes from red to yellow, whereas the indicator in tube B, the control, remains unchanged. It is therefore concluded that the green plant, *Cabomba*, has released CO_2 formed in its cells by respiration. The dark cover is used to prevent photosynthesis (see Chapter 16) from occurring.

If the experiment is repeated in bright light without covers over the tubes, the indicator in tube A is found to change from red to purple while that in tube B remains unchanged. It is therefore concluded that in bright light, the quantity of CO_2 taken in by the plant for photosynthesis exceeds the quantity of CO_2 given out as a result of respiration.

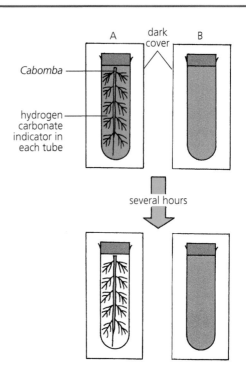

Figure 6.3 Respiration in a green plant

Testing Your Knowledge 1

1 Which sugar is the most common source of energy in living cells? (1)

2 a) What name is given to the series of enzyme-controlled reactions by which energy stored in sugar is released? (1)

   b) Where does this process occur in living organisms? (1)

3 Briefly describe the role of ATP molecules generated during the breakdown of glucose. (1)

4 The energy released during respiration is put to many uses by living cells. Name THREE of these uses. (3)

Biochemistry of respiration

Respiration is the process by which chemical energy is released during the breakdown of a food such as glucose. It occurs in every living cell and involves the generation of the high-energy compound ATP by a complex series of enzyme-controlled reactions.

Breakdown of glucose

Within a cell, the first stage of respiration is the breakdown of each molecule of glucose by a series of enzyme-controlled steps to form two molecules of **pyruvate** (pyruvic acid), as shown in Figure 6.4. This respiratory pathway releases enough energy to produce **two molecules of ATP** per molecule of glucose. It takes place in the cell's **cytoplasm** and does not need oxygen to be present.

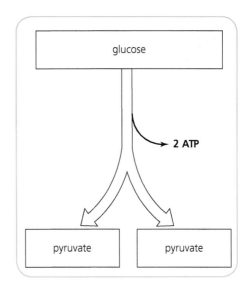

Figure 6.4 Breakdown of glucose

Aerobic respiration

When oxygen is present, aerobic respiration occurs in the cell's **mitochondria** (see Figures 6.5 and 6.6). The inner membrane of each mitochondrion is folded into many extensions that present a **large surface area** upon which the respiratory processes can take place. The extensions project into the central cavity, which contains essential enzymes. Cells such as sperm, liver, muscle and neurons in animals and companion cells (see page 96) in green plants that require much energy contain a relatively high number of mitochondria.

Fate of pyruvate

The respiratory pathway that takes place in mitochondria is illustrated in a simplified form in Figure 6.7. Each molecule of pyruvate (which diffuses into the mitochondrion from the cytoplasm) is broken down by many enzyme-controlled steps. This results in the formation of carbon dioxide and water, and the release of enough energy to yield a **large number of ATP molecules**. These high-energy ATP molecules make energy available for use by the cell. Aerobic respiration of glucose can be summarised as follows:

glucose + oxygen $\longrightarrow$ carbon dioxide + water + energy

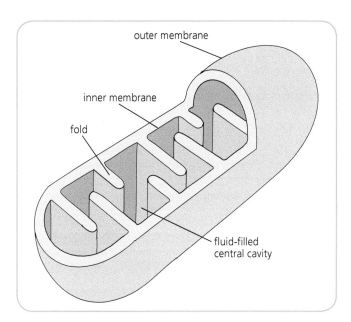

Figure 6.5 Mitochondrion

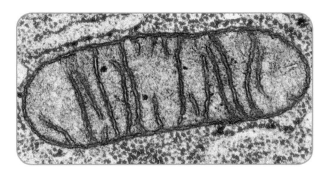

Figure 6.6 Electron micrograph of mitochondrion

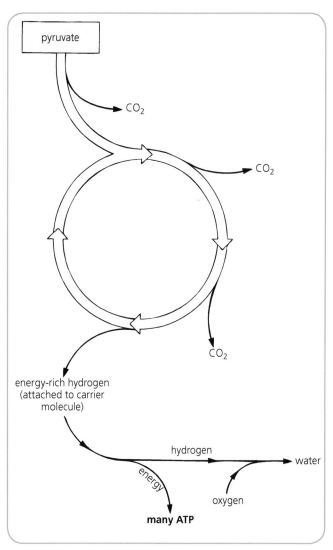

Figure 6.7 Aerobic breakdown of pyruvate

Respirometer

A **respirometer** is a piece of apparatus designed to measure **rate of respiration** (for example, volume of oxygen consumed per hour by a living organism). Figure 6.8 shows a simple respirometer. Carbon dioxide given out by the caterpillar is absorbed by the sodium hydroxide. Oxygen taken in by the caterpillar causes the volume of air in the enclosed system to decrease and the level of coloured liquid in the tube to rise. After one hour, the volume of air that must be injected (using the syringe) to return the coloured liquid to its initial level is equivalent to the volume of oxygen consumed per hour by the respiring caterpillar. (Note: Figure 6.8 does not show the control for this experiment. It would be an exact copy of the experiment except that it would lack a live caterpillar.)

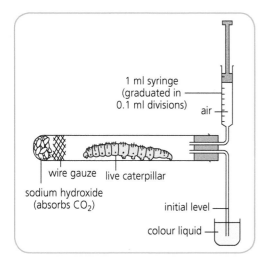

Figure 6.8

Investigating the effect of exercise on breathing rate

Breathing rate is the number of breaths taken per minute. Each breath involves breathing in and breathing out. With the aid of a stopwatch, the person times him/herself sitting quietly for 2 minutes to allow their breathing rate to settle at a steady pace. Then they measure their breathing rate for 1 minute while continuing to sit quietly.

Next the person exercises vigorously by running on the spot for 3 minutes. Then they measure their breathing rate immediately after stopping exercising. The person pools their results with other students and calculates **mean values** for breathing rate before and immediately after exercise. This procedure makes the results more **reliable**.

Exercise is found to make breathing rate increase. The person takes **more breaths** per minute than they do at rest. They also take **deeper** breaths. These changes to breathing help to ventilate the lungs more thoroughly during and after exercise. This increases the rate of gas exchange between the air in the air sacs and the blood flowing through the lungs. As a result, the person is able to gain the **oxygen** needed by respiring muscle cells to generate energy for movement. The increased breathing rate also enables the person to exhale more of the **carbon dioxide** formed by respiring muscle cells during exercise.

Investigation

Measuring rate of respiration in yeast

The experiment is shown in Figure 6.9.

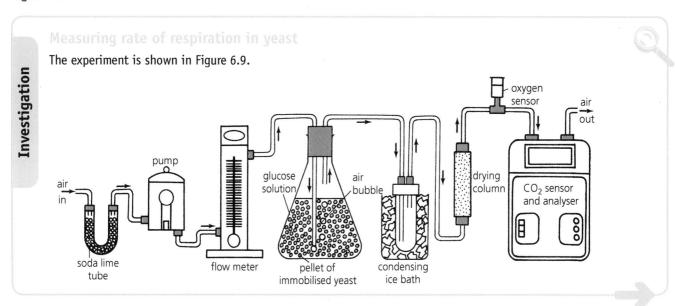

Figure 6.9 Measuring respiration rate

Table 6.2 gives the purpose of each piece of equipment. Live yeast cells, immobilised in gel pellets, are added to glucose solution in the conical flask at room temperature. Air is pumped through the system and the experiment is run for a set length of time (such as 30 minutes).

The computer software monitors the data from the sensors and displays the information on-screen. From these data the yeast's rate of respiration (for example, measured as **volume of oxygen consumed per unit time**) can be determined.

The immobilised yeast pellets are then rinsed and placed in a second conical flask of glucose solution in a water bath at 35 °C. This flask is given time to acclimatise to the temperature of the water bath. When the experiment is repeated at the higher temperature, the yeast's rate of respiration is found to increase.

Equipment	Purpose
soda lime tube (containing sodium hydroxide)	to absorb all carbon dioxide from incoming air so that its initial concentration is not a variable factor
air pump	to pump a continuous flow of air through the system
flow meter	to maintain the flow of air at a steady rate that is low enough for the carbon dioxide sensor to work
condensing bath and drying column	to remove water vapour from passing air since the sensors need air to be dry
oxygen probe (sensor)	to measure percentage oxygen concentration and send data to computer
carbon dioxide probe (sensor) and analyser	to measure carbon dioxide in parts per million and send data to computer

Table 6.2 Purposes of respirometer equipment

Fermentation pathway

This is the process by which a little energy is derived from the partial breakdown of glucose in the absence of oxygen. Breakdown of glucose begins as normal in the cell and **2 molecules of ATP** are formed per molecule of glucose. However, the pathway involving the complete breakdown of pyruvate to carbon dioxide and water cannot proceed in the absence of oxygen. Instead the pyruvate undergoes one of the following pathways.

Animal cells

During vigorous muscular activity when oxygen supply cannot meet demand, animal cells continue to respire in these anaerobic conditions by fermentation and **lactate** (lactic acid) is formed as in the following equation:

glucose ⟶ pyruvate ⟶ lactate + energy

Figure 6.10 represents this form of respiration in muscle cells in the human body when they are short of oxygen. As concentration of lactate builds up, it reduces the efficiency of the muscles. They may also suffer **muscle**

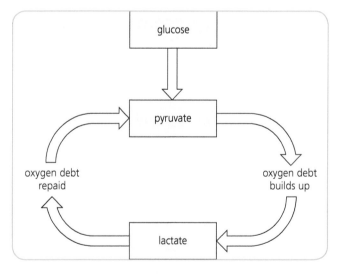

Figure 6.10 Fermentation in animal cells

fatigue. During this time, an **oxygen debt** builds up. The debt is repaid when oxygen becomes available during a rest period (see Figure 6.11) and energy, generated by aerobic respiration, is used to convert lactate back to pyruvate. The conversion of pyruvate to lactate in respiring animal cells is therefore described as a **reversible** process.

Figure 6.11 Repayment of oxygen debt

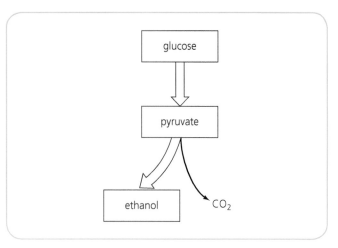

Figure 6.12 Fermentation in plant cells

Plant cells

Root cells of plants in waterlogged soil and yeast cells deprived of oxygen, for example during wine-making, are unable to respire aerobically. Such plant cells are dependent on **fermentation** (respiration in anaerobic conditions), as shown in Figure 6.12.

Since a molecule of carbon dioxide is lost each time a molecule of pyruvate is broken down to **ethanol**

(alcohol), fermentation in plant cells is an **irreversible** process. It is summarised by the following equation:

$$\text{glucose} \longrightarrow \text{pyruvate} \longrightarrow \text{ethanol} + \text{carbon dioxide} + \text{energy}$$

Comparison of aerobic respiration and fermentation

Table 6.3 compares the two forms of respiration with reference to one molecule of glucose.

	Aerobic respiration	Fermentation
Need for oxygen	oxygen always required	oxygen never required
Energy yield	efficient method releasing many ATP per molecule of glucose	inefficient method releasing 2 ATP per molecule of glucose
Degree of breakdown of glucose	glucose completely broken down	glucose partially broken down
End products	carbon dioxide and water	lactate in animal cells; ethanol and carbon dioxide in plant cells
Location of process in cell	begins in cytoplasm and is completed in mitochondria	occurs in cytoplasm

Table 6.3 Comparison of aerobic respiration and fermentation

Testing Your Knowledge 2

1 Write a word equation to summarise the complete breakdown of one molecule of glucose by aerobic respiration. (1)

2 a) Give the word equation that summarises anaerobic respiration in animal cells. (1)

   b) The following four statements refer to fermentation in human muscle tissue. Arrange them into the correct order. (1)

      A lactate builds up
      B oxygen supply becomes limited
      C muscles become less efficient
      D lactate is produced

   c) i) Is fermentation in animal cells reversible?
      ii) Justify your answer to i). (2)

3 a) Give the word equation that summarises fermentation in yeast cells. (1)

   b) i) Is this process reversible?
      ii) Justify your answer to i). (2)

4 Using the letters A and F, indicate whether each of the following statements refers to A (aerobic respiration) or F (fermentation). Some statements may need the use of both letters. (8)

   a) It involves the splitting of glucose into pyruvate.
   b) It yields many ATP from one molecule of glucose.
   c) It only yields 2 ATP from one molecule of glucose.
   d) It takes place in the cytoplasm only.
   e) It starts in the cytoplasm and is completed in the mitochondria.
   f) Its end products are CO_2 and water.
   g) Its end products are CO_2 and ethanol.
   h) It creates an oxygen debt in muscle cells.

What You Should Know Chapter 6

aerobic	energy	mitochondria
ATP	enzymes	oxygen
carbon	ethanol	pyruvate
cellular	fermentation	respiration
cytoplasm	lactate	two
dioxide	large number	water

1 The chemical _____ stored in glucose is released by a series of reactions controlled by _____. This biochemical process is called _____.

2 _____ is a high-energy compound that can release the energy required for _____ processes.

3 The first stage of the respiratory pathway is common to both _____ respiration and fermentation (anaerobic respiration). It involves the breakdown of glucose to _____.

4 In the presence of _____, aerobic respiration occurs. _____ and carbon _____ are formed and a _____ of molecules of ATP are produced per molecule of glucose.

5 In the absence of oxygen, _____ occurs. In animal cells, pyruvate is reversibly converted to _____. In plant cells, pyruvate is irreversibly converted to _____ and _____ dioxide. In each case only _____ molecules of ATP are produced per molecule of glucose.

6 Aerobic respiration begins in a cell's cytoplasm and is completed in its _____. Fermentation occurs in the _____ only.

Applying Your Knowledge and Skills Chapters 1–6

1 The rows in Table KS1.1 give triplets of features characteristic of certain cells.

Row	Triplet of features characteristic of cell		
1	mitochondria	nucleus	plasmids
2	cytoplasm	nucleus	mitochondria
3	large central vacuole	cell wall	mitochondria
4	cell wall	chloroplasts	circular chromosome
5	plasmids	circular chromosome	cell wall
6	ribosomes	plasmids	mitochondria

Table KS1.1

a) Which row refers correctly to both plant and animal cells? (1)
b) Which row is correct for plant cells only? (1)
c) Which row refers correctly to a bacterial cell? (1)

2 a) Figure KS1.1 shows three cells under the high power of a microscope. The diameter of the field of view is 0.03 mm. What is the average diameter of a cheek cell in micrometres? (1 millimetre = 1000 micrometres) (1)

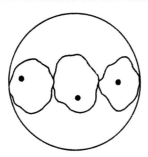

Figure KS1.1

b) A human red blood cell is 7 micrometres in diameter. Express this as a decimal fraction of a millimetre. (1)
c) If a cell is 0.008 millimetres long, what is its length in micrometres? (1)

3 Three identical cylinders of fresh turnip were immersed in the liquids shown in Figure KS1.2 for 24 hours. Each was then removed and held between forefinger and thumb as shown in Figure KS1.3.

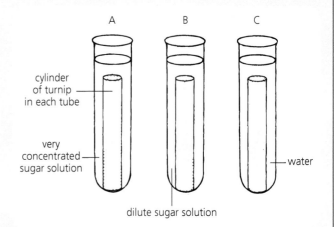

Figure KS1.2

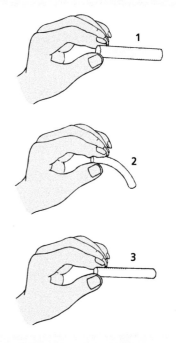

Figure KS1.3

a) Match numbers 1, 2 and 3 with letters A, B and C. (3)
b) Justify your choice in each case. (3)

4   Figure KS1.4 shows the direction of movement of two different substances through the cell membrane of an animal cell.

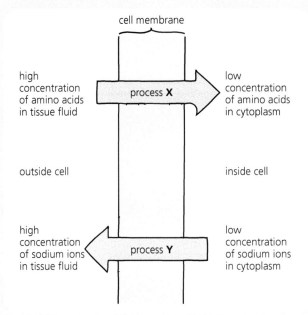

Figure KS1.4

a) i)   Name processes X and Y.
    ii) Which of these processes requires energy?
    iii) Which process will be unaffected by a decrease in oxygen concentration in the animal's environment? (4)
b) i)   Predict what will happen to the rate of process Y if the temperature of the cell is reduced to 4 °C for several hours.
    ii) Give a reason for your answer. (2)

5   Table KS1.2 shows a sample of Chargaff's data following the analysis of DNA extracted from several species.

Species	%G	%A	%C	%T
maize	22.9	27.1	22.9	27.1
wheat	22.7	27.3	**box P**	27.3
chicken	21.6	28.4	21.6	28.4
human	20.2	**box Q**	20.2	29.8

Table KS1.2

a) Study the data and calculate the figures that should have been entered in boxes P and Q. (2)
b) i)   State Chargaff's rules.
    ii) Do the data in the table support these rules?
    iii) Explain your answer to ii). (3)

6   Table KS1.3 gives the results from an experiment set up to investigate the effect of temperature on the action of a digestive enzyme.

Temperature (°C)	Mass of substrate broken down (mg/h)
0	0
5	1
10	4
15	8
20	14
25	22
30	28
35	31
40	32
45	29
50	18
55	0

Table KS1.3

a) Present the data as a line graph. (3)
b) i)   Explain what is meant by the term *optimum temperature*.
    ii) State the optimum temperature for the action of this enzyme. (2)
c) i)   By how many times was the rate of enzyme activity greater at 30 °C than at 20 °C?
    ii) Explain the difference in terms of rate of molecular movement and frequency of collisions between enzyme and substrate molecules at these two temperatures. (3)
d) Which rise in temperature of 5 °C brought about the biggest increase in rate of enzyme activity? (1)
e) At the temperature range 50–55 °C, the molecules are still gaining energy, so why does the reaction come to a halt at 55 °C? (1)
f) Predict the mass of substrate that would be broken down at 75 °C. (1)

7 The graph in Figure KS1.5 shows the effect of pH on the activity of three enzymes, X, Y and Z.

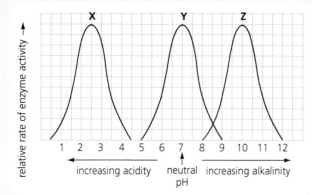

Figure KS1.5

a) State the working range of pH for each of the enzymes. (3)
b) What generalisation can be drawn about:
   i) the breadth of working range of pH of each of the enzymes?
   ii) the extent to which they all share the same actual pH working range? (2)

c) State the optimum pH for each of the enzymes. (3)
d) Suggest which of the enzymes would show optimum activity in:
   i) the human mouth
   ii) the human stomach. (2)

8 A transgenic animal is one that has had its genetic material modified deliberately by humans. This is often done by microinjecting the required gene(s) from one organism into the nuclei of the eggs of another species. Some people support this line of work and point out its potential benefits; others oppose it and claim that it is ethically wrong.

a) Find out what *ethics* means and try to describe it in your own words. (2)
b) Group the eight conversation bubbles shown in Figure KS1.6 into:
   i) those for the production of transgenic animals
   ii) those against the practice. (2)
c) i) Are you in favour of the development of transgenic animals?
   ii) Justify your answer to i).

A Blood anti-clotting factors needed by haemophiliacs can be obtained from the milk of transgenic cows, sheep and goats.

B The development of transgenic animals could change the direction of evolution for the worse.

C Flocks of transgenic sheep that produce more meat and more wool than normal would benefit the economy.

D Transgenic animals such as pigs could provide a reliable source of hearts needed for transplant surgery.

E If transgenic animals escaped into the wild, they could upset the balance of nature.

F The use of transgenic animals only considers the welfare of humans and not that of the animals involved.

G Transgenic animals are unnatural and could harbour some unknown diseases in the future.

H Transgenic animals that show disease symptoms can be used as disease models to find out which new treatments work.

Figure KS1.6

9 Read the passage and answer the questions that follow it.

During normal growth, plants produce ammonia as a by-product of their metabolism. This is normally rendered harmless as follows:

$$\text{ammonia} \xrightarrow{\text{enzyme action}} \text{end products}$$
$$\text{(toxic)} \qquad\qquad\qquad \text{(non-toxic)}$$

Glufosinate is an active ingredient in several brands of herbicide. It works by disrupting the enzyme action referred to in the above equation. As ammonia builds up, it destroys the plant's chloroplasts.

Some plants have been genetically modified for resistance to glufosinate in the following way.

A gene that deactivates glufosinate was found to occur in species of *Streptomyces* bacteria. This gene was transferred to a type of soil bacterium that normally infects plant roots and injects its DNA into the plants. The genetically modified (GM) version of this bacterium was used to deliberately infect certain crop plants and introduce the gene for resistance to glufosinate into them. By this means, some of the crop plants were successfully transformed into GM plants resistant to herbicide containing glufosinate. When a crop of these resistant plants is sprayed with weedkiller, the crop survives but the weeds growing beside it die. This has led to a reduction in use of herbicide from a mass of 7 kg/ha to as little as 0.7 kg/ha in many cases. (ha = 1 hectare = 100 acres)

a) Give the meaning of the term *herbicide*. (1)
b) Explain why the destruction of chloroplasts is fatal to a plant. (1)
c) Describe the means by which a crop plant resistant to herbicide is produced. (3)
d) Calculate the percentage reduction in mass of herbicide used as a result of growing GM crops. (1)
e) *'Resistance to herbicide occurs naturally in the environment.'* Justify this statement using information from the passage. (1)

10 Figure KS1.7 overleaf shows six versions of the apparatus set up during an investigation to compare the energy content of two different foods.

a) Explain fully why a valid comparison cannot be made between set-ups:
   i)   A and B
   ii) A and C
   iii) A and E. (6)
b) Describe THREE changes that would need to be made to set-up C so that a valid comparison could be made between it and D. (3)
c) The experiment was carried out using set-ups B and F and the results in Table KS1.4 were obtained.

Food	Water temperature (°C)		Temperature rise (°C)	Energy content (kJ)
	at start	at finish		
peanut	25	47		
glucose	26	38		

Table KS1.4

   i)   Calculate the energy content of the two foodstuffs in kilojoules per gram using the formula:

   energy released (kJ) = 4.2MT/1000

   where M = mass of water (g) and T = rise in temperature (°C)
   (Note: 1 cm^3 of water has a mass of 1 g.)

   ii) A data book was found to give the energy content per gram for peanut as 25 kJ and for glucose as 17 kJ, measured using a food calorimeter. The fact that the results in Table KS1.4 are much lower than these is accounted for by the fact that food burned in air under a beaker loses heat to the surroundings and does not burn completely to ashes. Study the diagram of the food calorimeter on page 37 and explain how use of it largely overcomes these two shortcomings. (4)

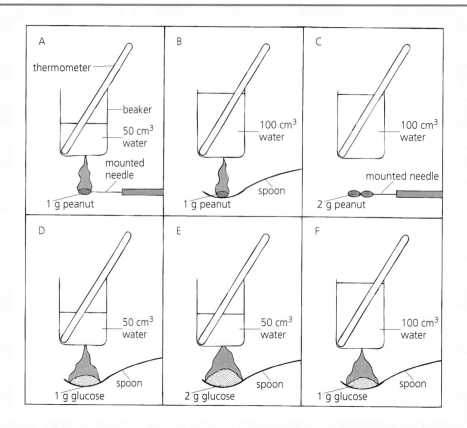

Figure KS1.7

11 The information in Table KS1.5 refers to hydrogen carbonate indicator. The experiment in Figure KS1.8 was run for 24 hours using green cells of a fast-growing alga.

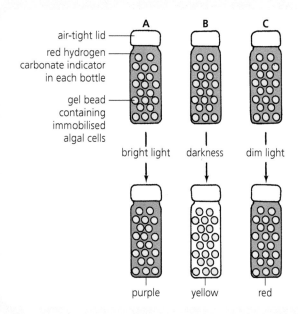

Figure KS1.8

Relative carbon dioxide concentration of indicator	Colour of indicator	pH of indicator
very high	yellow	7.6
high	orange	8.0
medium	red	8.4
low	reddish-purple	8.8
very low	purple	9.2

Table KS1.5

a) What relationship exists between the pH of hydrogen carbonate indicator and its relative concentration of carbon dioxide? (1)

b) Account fully for the final colour of the indicator in each of bottles A, B and C. (6)

c) i) If the same mass of algae immobilised in smaller gel pellets was used, what effect would this have on the time required for the colour change to occur in A?

ii) Explain your answer to i). (2)

(Since this group of questions does not include examples of every type of question found in SQA exams, it is recommended that students also make use of past exam papers to aid learning and revision.)

2

Multicellular Organisms

7 Producing new cells

Growth is the irreversible increase in the dry mass of an organism. It is normally accompanied by an increase in cell number as a result of repeated cell division.

Cell division

Figure 7.1 shows cell division in an animal cell. The nucleus divides first and the two daughter nuclei separate. Next the cytoplasm becomes pinched off between the two nuclei, forming two daughter cells.

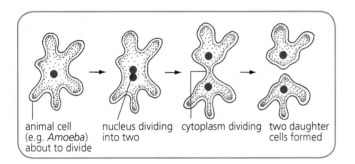

animal cell (e.g. *Amoeba*) about to divide | nucleus dividing into two | cytoplasm dividing | two daughter cells formed

Figure 7.1 Cell division in an animal cell

Division in a plant cell is shown in Figure 7.2. The cytoplasm cannot become pinched off in the middle because of the presence of the cell wall. Instead, nuclear division is followed by the laying down of a new cell wall between the daughter cells.

Chromosomes

Chromosomes (see Figure 7.3) are threadlike structures found inside the nucleus of plant and animal cells. Each chromosome carries information that is necessary for the development of the cell and the survival of the living organism to which it belongs.

Chromosome complement

Each species of plant and animal has a definite and characteristic number of chromosomes called its chromosome complement present in each cell.

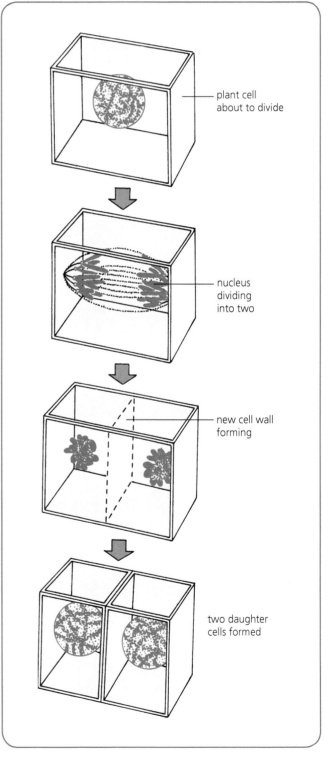

plant cell about to divide

nucleus dividing into two

new cell wall forming

two daughter cells formed

Figure 7.2 Cell division in a plant cell

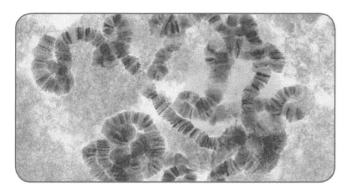

Figure 7.3 Chromosomes

Every normal human body cell, for example, contains 46 chromosomes as 23 pairs (see Figure 7.4). Such a cell, which has two matching sets of chromosomes, is described as **diploid**. Table 7.1 gives some examples of diploid chromosome complements. Figure 7.5 shows the diploid chromosome complement of the fruit fly, which is composed of two identical sets of four chromosomes.

Species of living thing	Diploid chromosome complement
onion	16
cabbage	18
rice	24
fruit fly	8
frog	24
cat	38
human	46
horse	66

Table 7.1 Diploid chromosome complements

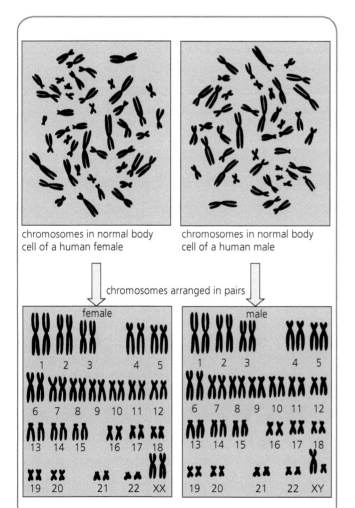

Figure 7.4 Matching sets of chromosomes

Figure 7.5 Diploid chromosome complement of fruit fly

Every normal human sex cell such as a sperm contains 23 unpaired chromosomes. Such a cell, which has a single set of chromosomes, is described as **haploid**. Diploid cells are often represented by the symbol $2n$ because they have two sets of chromosomes. Haploid cells, which have a single set of chromosomes, are represented by the symbol n (where n equals the number of different chromosomes).

Mitosis

Mitosis is the process by which the **nucleus** divides into two daughter nuclei, each of which receives exactly the same number of chromosomes as were present in the original nucleus. Figure 7.6 shows the sequence of events that occurs during mitosis. Figure 7.7 shows a microscopic view of some of the stages in root tip cells that are dividing.

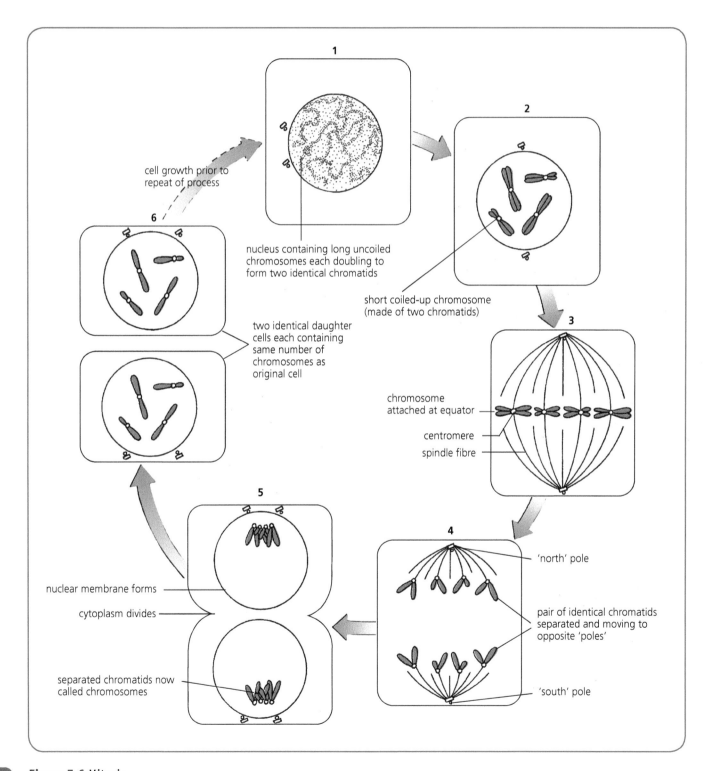

Figure 7.6 Mitosis

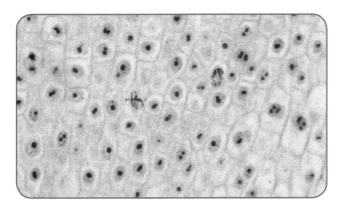

Figure 7.7 Mitosis in root tip cells

As each chromosome becomes shorter and thicker, it is seen to be a double thread. Each thread is called a **chromatid.** The two chromatids have resulted from the exact **replication** (reproduction) of a chromosome. They are held together by a **centromere.** The nuclear membrane disappears and a spindle made of **spindle fibres** forms. Each chromosome becomes attached by its centromere to one of the spindle fibres at the **equator.** Next, each centromere splits and one chromatid from each pair moves to the 'north pole' and one to the 'south pole'. Finally a nuclear membrane forms around each group of chromatids (now regarded as chromosomes) and nuclear division (mitosis) is complete.

Mitosis is followed by division of the cytoplasm to form **two identical daughter cells.** Since each of the two cells produced has received a complete set of chromosomes, each new cell contains a copy of exactly the same information. Each cell now undergoes a period of cell growth. During this time the chromosomes in each nucleus cannot be seen and sometimes this stage (called interphase) is referred to as the resting period. However, the chromosomes are not resting. During this time, each is completely uncoiled and undergoing **replication.** By this means, an exact copy of each chromosome (a chromatid) is formed in preparation for the next nuclear division.

Maintenance of diploid chromosome complement

Chromosomes provide the main source of genetic information typical of a particular species of living thing. It is essential that each cell formed as a result of mitosis receives a full complement of chromosomes so that during growth and development, the cells of the multicellular organism will be able to provide the animal or plant with the characteristics of its species. This **continuity of diploid chromosome complement** from cell to cell is maintained by **mitosis.** Mitosis is required for growth and for the replacement of damaged cells during tissue repair.

Related Topic

Length of time

The time taken by a cell to go through the entire cycle of cell division from interphase to interphase varies from species to species. An onion root tip cell takes 22 hours to complete the full cycle at room temperature. Of this time, mitosis (nuclear division) only takes 90 minutes. The remainder of the time is spent on cell growth and formation of chromatids.

Related Topic

Growth curves

Measuring growth

Growth is usually investigated by measuring changes in an organism's fresh mass, height or cell number over a period of time.

Growth curve

Table 7.2 overleaf records the height of a plant shoot measured over a period of 16 days. When graphed, the results for total height of shoot give a **growth curve** as shown in Figure 7.8.

Careful study of Figure 7.8 and the data in Table 7.2 reveals that during the first 4 days of growth, each daily increase in height was greater than that of the previous day. This phase is called the period of **accelerating growth.** During days 4 to 8, the daily increase in height remained constant. This phase is called the period of **steady rapid growth.** During days 8 to 12, growth continued but each day's increase in height was less than that of the previous day. This phase is called the **period of decelerating growth.** During days 12 to 16, the shoot failed to gain further height. It had entered the **period of no growth.** This pattern of growth, with minor variations, is common to almost all living things. Figure 7.9 shows a generalised version of the growth curve and its four phases.

Time from start (days)	Total height of shoot (mm)	Daily increase in height of shoot (mm)	Growth phase
start	0	-	period of accelerating growth
1	1	1	
2	3	2	
3	7	4	
4	15	8	
5	23	8	period of steady rapid growth
6	31	8	
7	39	8	
8	47	8	
9	51	4	period of decelerating growth
10	53	2	
11	54	1	
12	54	0	
13	54	0	period of no growth
14	54	0	
15	54	0	
16	54	0	

Table 7.2 Growth of a plant shoot

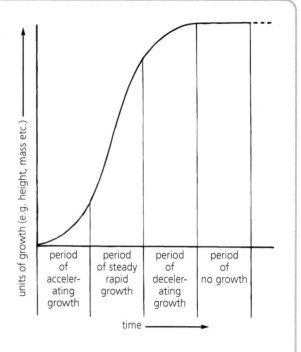

Figure 7.9 Growth curve

Human growth curve

Growth in humans, as indicated by change in body mass, gives a similar curve but it has two phases of rapid growth (called **growth spurts**) instead of one (see Figure 7.10). The first occurs during the two years following birth; the second occurs at puberty after a long period of steady growth. On average, a human male reaches puberty later and attains a larger adult size than a human female.

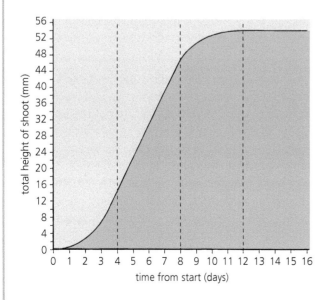

Figure 7.8 Graph of growth of height of plant shoot

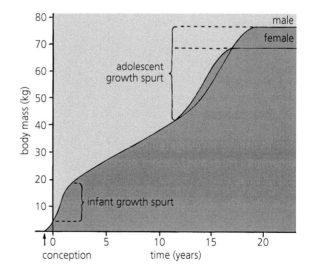

Figure 7.10 Human growth curves

Testing Your Knowledge 1

1 a) State the main events that occur during cell division in an animal cell. (2)
  b) In what way does cell division in a plant differ from that in an animal? (1)
2 Which cell structure controls all cell activities including cell division? (1)
3 Arrange the stages of mitosis shown in Figure 7.11 into the correct sequence, beginning with B. (1)

4 Briefly describe the process of mitosis using all of the following terms: *centromere, chromatid, equator, spindle fibre.* (4)
5 Why is it important that the diploid chromosome complement of daughter cells in a multicellular organism is maintained? (2)

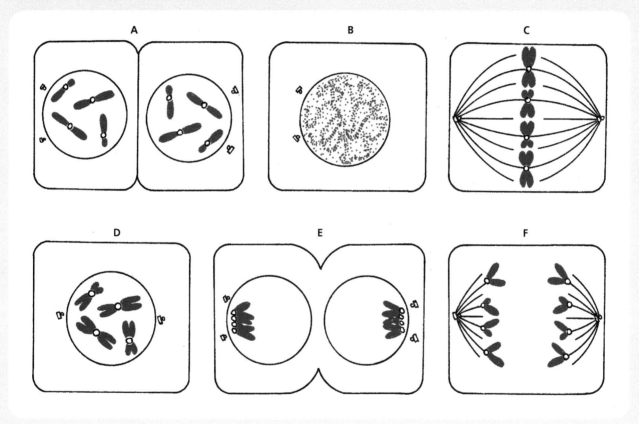

Figure 7.11

Stem cells

Stem cells (see Figure 7.12) are unspecialised animal cells involved in growth and repair. They are able to:

- divide in order to self-renew
- reproduce themselves indefinitely by repeated mitosis and cell division while remaining unspecialised
- develop into various types of specialised cell when required, for example, to replace cells that have come to the end of their life or have been damaged or lost.

Two types of stem cell

An early human embryo (see Figure 7.13) contains **embryonic stem cells**. These are capable of developing into all cell types found in the human body. They even have the potential to regenerate an entire organ from a few cells.

Throughout life a fully formed human possesses **tissue stem cells** at various locations such as bone marrow, blood and skin. These cells have a more limited potential than embryonic stem cells. Under natural conditions,

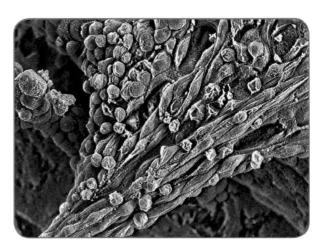

Figure 7.12 Stem cells

tissue stem cells are only able to replenish the supply of one or more types of specialised cell closely related to the tissue in which they are found. Tissue stem cells in red bone marrow, for example, can only give rise to new blood cells while those in skin can only produce more skin cells.

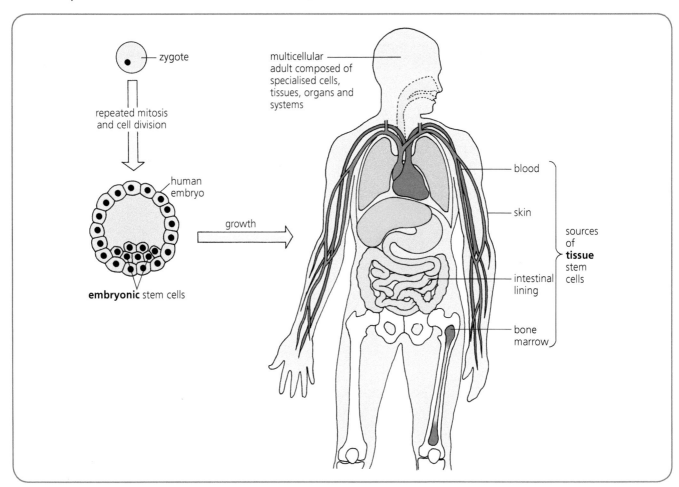

Figure 7.13 Two types of stem cell

Potential uses of stem cells

Human stem cells can be grown in cultures provided that they are given optimum conditions and a supply of certain key growth factors. In recent times, stem cells have been successfully put to use in a variety of ways, as in the following examples.

Bone marrow transplantation

Some forms of cancer of the blood such as leukaemia result from the uncontrolled increase in number of white blood cells. One form of treatment involves destruction of the patient's cancerous bone marrow cells and their replacement with a transplant of normal, blood-forming stem cells from a suitable donor (see Figure 7.14).

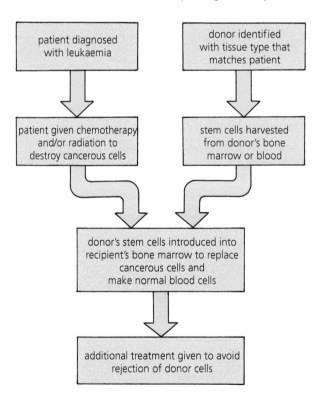

Figure 7.14 Bone marrow transplantation

Skin graft

In a traditional skin graft, a relatively large section of skin is removed from a region of the person's body and grafted to the site of injury. This means that the person has two parts of their body that need careful treatment and time to heal.

A skin graft using stem cells only requires a **small sample** of skin to be taken to obtain stem cells. Therefore the site needs much less healing time and suffers minimum scarring. The sample is normally taken from an area close to, and similar in structure to, the site of injury. Enzymes are used to isolate and loosen the stem cells, which are then cultured. Once a **suspension of new stem cells** has developed, they are sprayed over the damaged area (see Figure 7.15) to bring about regeneration of the missing skin.

Figure 7.15 Spray-on stem cell skin graft

Cornea repair

In recent years scientists have shown that **corneal damage** by chemical burning can be successfully treated using stem cell tissue. This can be grown from the patient's own stem cells located at the edge of the cornea. In many cases the person's eyesight can be restored following grafting of the stem cell tissue from the healthy eye to the surface of the damaged eye (see Figure 7.16). Since the skin graft and cornea repair techniques use the affected individual's own cells, there is no risk of the transplanted tissue being rejected.

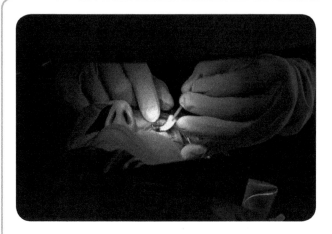

Figure 7.16 Repair of cornea using stem cells

Future therapeutic potential of stem cells

Embryonic stem cells are able to differentiate into any type of cell in the body. Therefore they are believed to have the potential to provide treatments in the future for a wide range of disorders and degenerative conditions such as **diabetes**, **Parkinson's disease** and **Alzheimer's disease** that traditional medicine has been unable to cure. Already scientists have managed to generate nerve cells from embryonic stem cells in culture. It is hoped that this work will eventually be translated into effective therapies to treat disorders such as multiple sclerosis. However, the use of embryonic stem cells raises questions of **ethics** (see below).

Related Topic

Ethical issues associated with use of stem cells

Ethics refers to the moral values and rules that ought to govern human conduct. The use of stem cells raises several **ethical issues**.

Embryonic stem cells

At present, the creation of a continuous culture of embryonic stem cells (a **stem line**) for research purposes makes use of cells taken from a human embryo of no more than 14 days' growth and development. Some people consider this practice to be unethical because it results in the destruction of the embryo. It is interesting to note that normally the embryos used to obtain stem cells were generated for *in vitro* (IV) fertilisation but were additional to requirements. They would have been destroyed eventually as a matter of course.

People on one side of the debate believe that an embryo is a human *person* and that fatally extracting stem cells from it is equivalent to committing murder. People on the other side of the debate believe that the ball of cells that makes up a 14-day-old embryo is not yet a person and that removing stem cells from it for research purposes is morally acceptable.

Nuclear transfer technique

One form of this technique (see Figure 7.17) involves removing the nucleus from an egg cell (e.g. from a cow) and replacing it with a nucleus from a donor cell (e.g. a skin

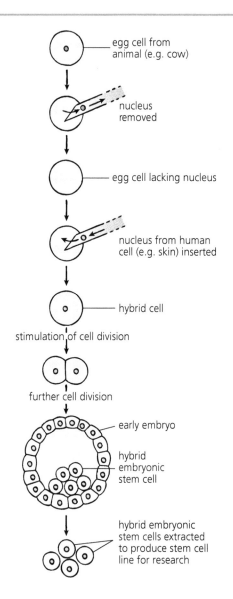

egg cell from animal (e.g. cow)

nucleus removed

egg cell lacking nucleus

nucleus from human cell (e.g. skin) inserted

hybrid cell

stimulation of cell division

further cell division

early embryo

hybrid embryonic stem cell

hybrid embryonic stem cells extracted to produce stem cell line for research

Figure 7.17 Nuclear transfer technique

cell from a human). Once the hybrid cell formed begins to divide, stem cells are extracted from it after five days and these are used for research. However, these cells are not 100% human and are only allowed to be used for research.

Some people feel that it is unethical to mix materials from human cells with those of another species even if guarantees are given that use of the hybrid cells will be carefully restricted. Other people support the production of these

hybrid cells because it helps to relieve the shortage of human embryonic cells available for research purposes.

Tissue stem cells

The use of tissue stem cells is not normally regarded as unethical because the cells have been obtained, with permission, from the tissue of an adult and their use does not cause the destruction of a human embryo.

Levels of organisation

Cell

A **cell** is often described as the basic unit of life because it is the smallest unit that can lead an independent life. The body of a multicellular organism consists of a large number of cells. It would be inefficient for every one of these cells to carry out every function essential for the maintenance of life. Instead, a **division of labour** is achieved by the cells being organised into tissues and the tissues being united to form organs.

Tissue

A **tissue** is a group of cells specialised to perform a particular function (or functions). Figures 7.18 and

7.19 (overleaf) illustrate a selection of tissues from two multicellular organisms. Tables 7.3 and 7.4 (on pages 60 and 61) describe how the structure of each of the cells is related to its function.

Organ

An **organ** is a structure composed of several different tissues co-ordinated to perform one or more functions. Tables 7.5 and 7.6 (on page 61) give some examples of organs in the human body and in a flowering plant.

System

A group of related tissues and organs (such as blood, heart, arteries, capillaries and veins) make up a **system** (such as the circulatory system).

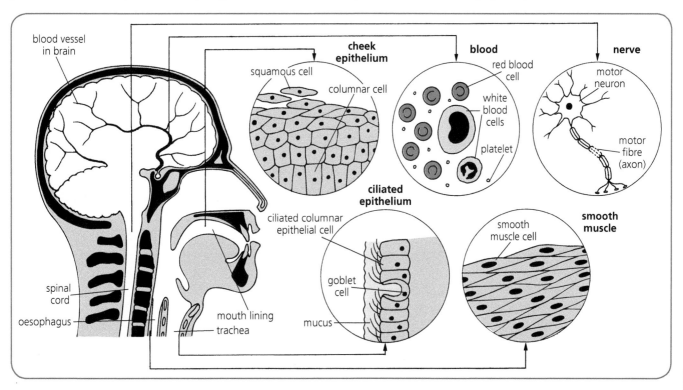

Figure 7.18 Human tissues and cells (not drawn to scale)

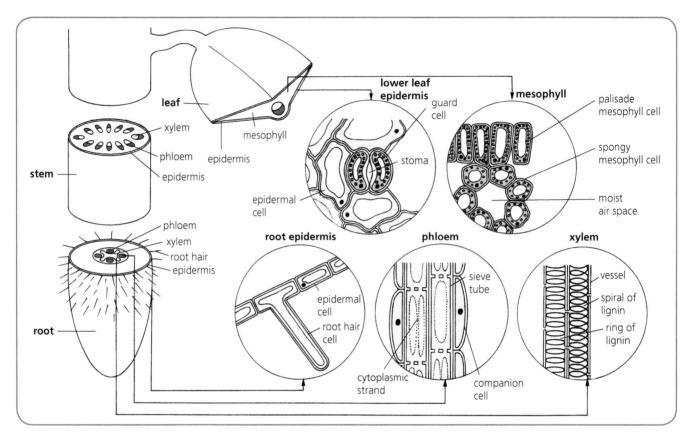

Figure 7.19 Plant tissues and cells (not drawn to scale)

Tissue	Cell type	Specialised structural features	Function
cheek epithelium	epithelial cell	flat, irregular shape (allowing cells to form a loose covering layer constantly replaced during wear and tear)	protection of mouth lining
blood	red blood cell	small size and biconcave shape present a relatively large surface area; rich supply of haemoglobin present	uptake and transport of oxygen to living cells
	white blood cell	able to change shape; sacs of microbe-digesting enzymes present in some types	destruction of invading microbes
nerve	motor neuron	axon (long insulated extension of cytoplasm)	transmission of nerve impulses
ciliated epithelium	goblet cell	cup shape; able to produce mucus	secretion of mucus, which traps dirt and germs
	ciliated epithelial cell	hair-like cilia that beat upwards	sweeping of dirty mucus up and away from lungs
smooth muscle	smooth muscle cell	spindle shape (allowing cells to form sheets capable of contraction)	movement of food down oesophagus by peristalsis

Table 7.3 Structure of animal cells in relation to function

Tissue	Cell type	Specialised structural features	Function
lower leaf epidermis	epidermal cell	irregular shape (allowing cells to fit like a jigsaw into a strong layer)	protection
	guard cell	sausage shape; thick inner cell wall facing stoma; chloroplasts present	control of gaseous exchange by changing shape resulting in opening and closing of stomata
mesophyll	palisade mesophyll	chloroplasts present; columnar shape (allows densely packed green layer to be presented to light)	primary region of light absorption and photosynthesis
	spongy mesophyll	'round' shape allows loose arrangement in contact with moist air spaces for absorption of carbon dioxide	secondary region of photosynthesis
phloem	sieve tube	sieve plates and continuous stream of cytoplasmic strands (but no nucleus)	transport of soluble carbohydrates up and down the plant
	companion cell	large nucleus in relation to cell size	control of sieve tube functions
xylem	vessel	hollow tube; wall strengthened by lignin; lignin deposited as rings or spirals allowing expansion and contraction	support and water transport up the plant
root epidermis	epidermal cell	box-like shape allowing cells to fit together like a brick wall	protection
	root hair cell	long extension of an epidermal cell presenting a large surface area in contact with soil solution	absorption of water and mineral salts

Table 7.4 Structure of plant cells in relation to function

Organ	Function(s)
stomach	churning and partial digestion of food
heart	pumping of blood
kidney	maintenance of water balance and removal of soluble wastes
lung	exchange of respiratory gases
skin	protection, temperature regulation and sensitivity

Table 7.5 Organs in the human body

Organ	Function(s)
root	anchorage and absorption of water
stem	support and transport of water and soluble food
leaf	photosynthesis and gaseous exchange
flower	production of seeds

Table 7.6 Organs in a flowering plant

Related Information

Organism as an integrated whole

For a complex multicellular organism to lead an independent life, all of its cells, tissues, organs and systems must operate in close co-ordination with each playing its particular role harmoniously as part of an **integrated whole**.

In the rat shown in Figure 7.20, for example, the chest cavity contains organs such as the lungs and trachea (parts of the **respiratory** system) and the heart and blood vessels (parts of the **circulatory** system). The chest cavity is protected by ribs made of bony tissue and moved during breathing by intercostal muscles.

The abdominal cavity contains the stomach and intestines (organs belonging to the **digestive** system), the kidneys and bladder (**excretory** system) and, in females, the ovaries, oviducts and uterus (**reproductive** system). In the male, the testes are found outside the cavity in the scrotal sac. The animal's whole body is moved by skeletal muscles supported by the bony skeleton and co-ordinated by the **nervous** system.

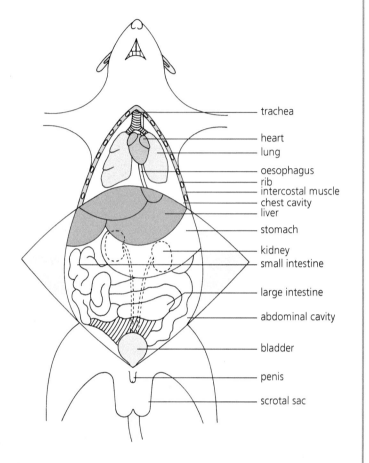

Figure 7.20 Organism as an integrated whole

Testing Your Knowledge 2

1 a) What is a *stem cell*? (2)
   b) Identify TWO places in the human body where stem cells are found. (2)

2 Describe a structural feature possessed by some white blood cells that enables them to destroy disease-causing bacteria. (1)

3 a) i) State the function of cheek epithelial cells.
      ii) Explain how these cells are suited to perform this function. (2)
   b) What role is played by the axon of a motor neuron? (1)

4 a) i) State TWO functions of xylem tissue.
      ii) Choose one of these functions and describe how a xylem vessel is structurally suited to perform it. (3)
   b) Identify TWO types of leaf cell capable of photosynthesis. (2)

5 Define the terms *tissue* and *organ*. (2)

What You Should Know Chapter 7

animal	equator	specialised
centromere	fibre	stem
chromatids	mitosis	system
chromosome	nucleus	tissues
complement	organs	unspecialised
diploid	pole	wall

1 During cell division, the _____ divides first followed by the cytoplasm. In a plant cell, a new cell _____ is laid down between the two daughter cells formed.

2 Every species has a characteristic number of chromosomes called its chromosome _____. Normally this takes the form of two matching sets of chromosomes and the cell that contains them is said to be _____.

3 The process by which the nucleus of a diploid cell divides into two nuclei, each of which receives a copy of the diploid _____ complement, is called _____.

4 During mitosis each chromosome, consisting of two identical _____, becomes attached by its _____ to a spindle _____ at the cell's _____. One chromatid from each chromosome moves to the north pole and one to the south _____ forming two nuclei.

5 Stem cells are _____ cells involved in the growth and repair of a multicellular _____. Division of _____ cells produces cells that have the potential to become different types of specialised cell.

6 In a multicellular organism, groups of _____ cells are organised as _____, which are grouped together to form _____. Groups of related tissues and organs make up a _____.

8 Control and communication

Nervous control

The body of a multicellular animal, such as a human, is composed of many different types of cells, tissues and organs. The majority of these structures are specialised to perform particular functions but they do not operate independently of one another. The body can survive and function effectively only if all of its parts work in close co-operation with one another. The internal communication needed to bring this about is provided by the **nervous system**.

In humans, the nervous system (see Figure 8.1) is composed of three parts – the **brain**, the **spinal cord** and the **nerves**. The brain and spinal cord make up the **central nervous system (CNS)**. The CNS is connected to all parts of the body by the nerves, which lead to and from all organs and systems. This arrangement ensures that all parts work together as a **co-ordinated whole** with the brain exerting overall control.

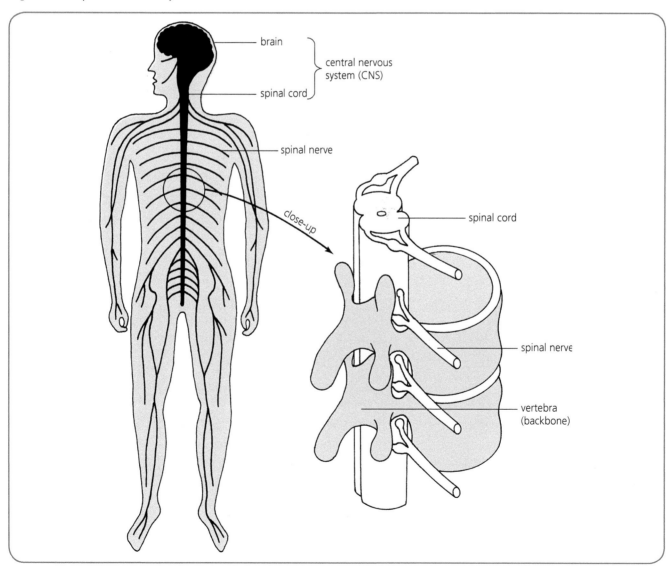

Figure 8.1 Human nervous system

Brain

The brain consists of several different regions, as shown in Figures 8.2 and 8.3. The **medulla** controls the rate of breathing and heart beat. The **cerebellum** controls balance and muscular co-ordination. The largest part of the brain is called the **cerebrum**. It is responsible for mental processes such as memory, reasoning, imagination, conscious thought and intelligence.

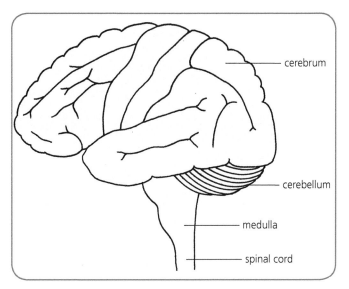

Figure 8.2 Human brain

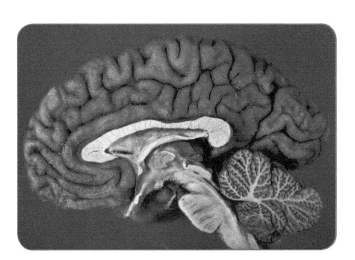

Figure 8.3 Human brain in longitudinal section

Flow of information

Figure 8.4 shows how one set of nerves (sensory) carries information from the body's **receptors** to the CNS and another set of nerves (motor) carries nerve impulses from the CNS to the body's **effectors**. Nerve messages arriving from receptors in the sense organs keep the CNS (particularly the brain) informed about all aspects of the body and its surroundings. The CNS sorts out all of this information and stores some of it. When the body is required to make a response, the CNS sends out nerve messages to the appropriate effector, such as a muscle or a gland. A response from a muscle is normally more rapid than a response from a gland.

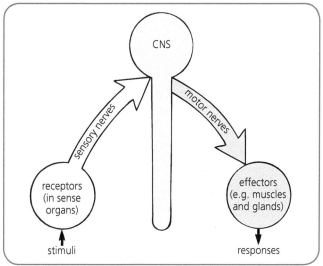

Figure 8.4 Flow of information

Neurons and reflex arc

The nervous system is made up of nerve cells called **neurons**. A neuron consists of a cell body attached to nerve fibres (see Figures 8.5 below and 8.6 overleaf).

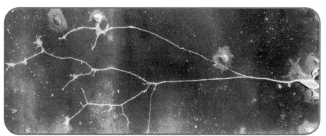

Figure 8.5 Neuron

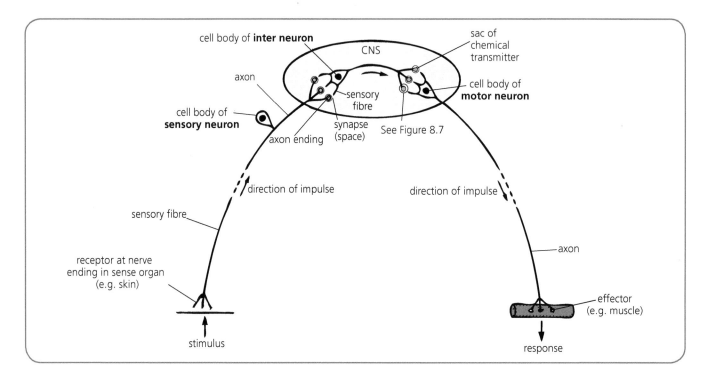

Figure 8.6 Reflex arc

An **electrical impulse** is carried towards the cell body of a neuron by a **sensory fibre** and away from it by an **axon fibre**. The simple arrangement of three different types of neuron shown in Figure 8.6 is called a **reflex arc**.

A tiny space called a **synapse** occurs between the axon ending of one neuron and the sensory fibre of the next. When a nerve impulse arrives, the ending of the axon branch (see Figure 8.7) releases a chemical. This diffuses across the space between the two neurons and triggers off an impulse in the sensory fibre of the next neuron in the arc. By this means, the chemical brings about the transfer of a message between two neurons.

Information collected by receptors is passed by **sensory** neurons to the CNS to be processed. Nerve impulses are then transmitted via **inter** neurons to **motor** neurons, which enable a response to be made by an effector. This may take the form of a slow response by a gland or a more rapid response by a muscle.

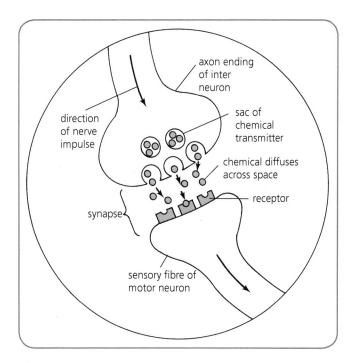

Figure 8.7 Synapse

Pain receptors

The human body has many receptors that detect information about the body's internal and external environments. This information is normally sent to the brain as electrical impulses via the nervous system, which functions as a vast collection of interrelated pathways of communication. Receptors are present in all sense organs. The skin, for example, contains receptors for touch, heat, cold, pressure and pain (see Figure 8.8).

When the brain receives nerve impulses from **pain receptors**, the affected person perceives the **sensation of pain**. Pain is of survival value because it makes the person aware of the danger. It gives them an opportunity to adopt some corrective method of behaviour such as moving a hand away from a source of intense heat before it becomes seriously injured.

People who are born without the ability to perceive pain are more vulnerable to injury because they lack this essential warning system. Such a person could even die from a severe condition such as a ruptured appendix because they would be unaware that it had taken place.

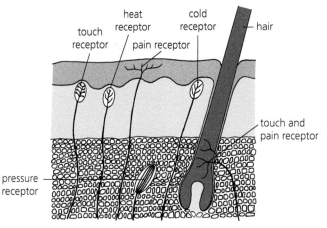

Figure 8.8 Receptors in human skin

1 a) i) Name THREE parts of the human nervous system.
   ii) Which of these make up the CNS (central nervous system)? (2)
   b) Briefly describe the role played by the nervous system. (1)

2 a) Which region of the brain is concerned with learning and memory? (1)
   b) Name TWO other parts of the human brain and for each state its function. (4)

3 The following structures are involved in the transmission of a nerve impulse. Arrange them in the correct order: *motor nerve, sense organ, muscle, CNS, sensory nerve.* (1)

Reflex action

The transmission of a nerve impulse through a reflex arc results in a **reflex action**. A reflex action is a rapid, automatic, involuntary response to a stimulus. Figure 8.9 overleaf shows an example of a reflex action called limb withdrawal. When the back of the hand accidentally comes in contact with intense heat, this **stimulus** is picked up by pain receptors in the skin (1) and an electrical impulse is immediately sent up the fibre of the sensory neuron (2). In the grey matter of the spinal cord, the impulse crosses the first synapse (3) and passes through the inter (relay) neuron (4). Once across the second synapse, the nerve impulse is picked up by the motor neuron (5) and quickly conducted to the axon endings (6), which are in close contact with the flexor muscle of the arm. Here a chemical is released that brings about muscular contraction (the **response**) making the arm bend and move out of harm's way.

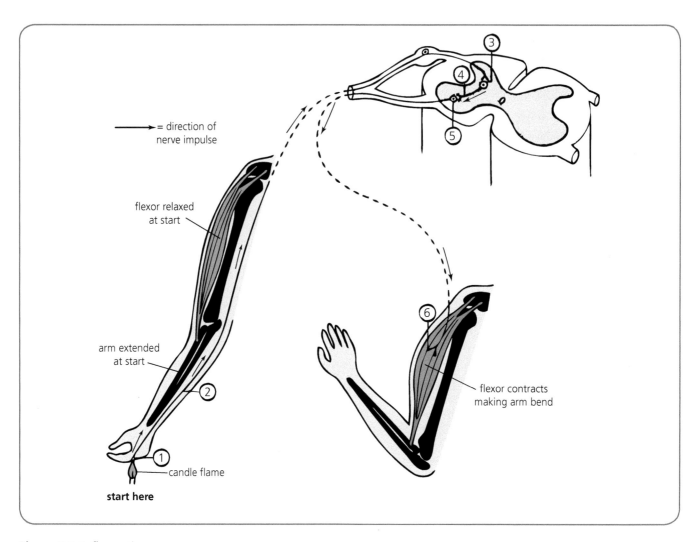

Figure 8.9 Reflex action

Eye iris reflex

The eyes of a volunteer are carefully observed in daylight and the 'normal' state of the pupils noted. Next, the person's eyes are kept closed and covered (to create darkness) for several minutes. On being opened, the eyes are immediately observed. The pupils, which have become enlarged in diameter in darkness (see Figure 8.10), are seen to quickly return to their normal size.

The person's eyes are given a few minutes to become accustomed to normal daylight. Then, when a bright lamp is held close to the person's eyes for 5 seconds, the pupils are seen to become reduced in diameter.

The iris of the eye contains two sets of antagonistic muscles (see Figure 8.11). These control the size of the pupil by bringing about an appropriate reflex action depending on the intensity of light to which the eye is exposed.

Protection

The activity of the iris muscles regulates the quantity of light able to enter the eye through the pupil. This is of protective value to the body for the following reasons. Maximum light is admitted in poorly lit conditions enabling the person to see and avoid harmful objects; minimum light is admitted in very brightly lit conditions preventing damage to the retina (the layer at the back of the eye composed of light-sensitive receptor cells).

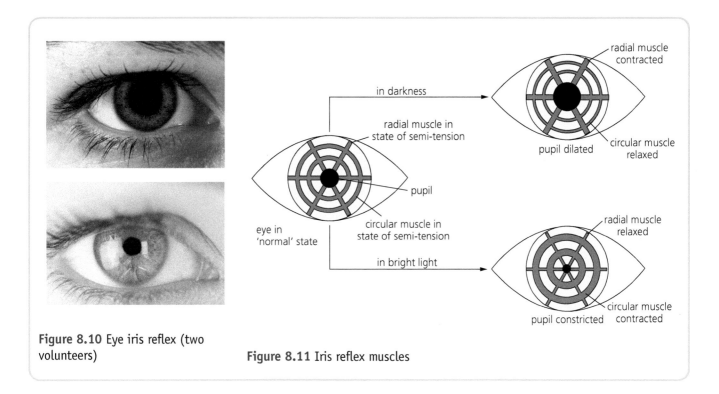

Figure 8.10 Eye iris reflex (two volunteers)

Figure 8.11 Iris reflex muscles

Protective role

Reflex actions **protect** the body from damage. Since they do not need conscious thought by the brain, many reflex actions may still be performed for a short period by an animal whose brain has been destroyed.

Hormonal control

In addition to the control exerted by the nervous system, further co-ordination of the workings of the human body is brought about by **hormones**. Hormones are **chemical messengers** secreted directly into the bloodstream by **endocrine** (ductless) glands (see Figure 8.12).

Hormones stimulate specific **target tissues**. The cells on the surface of a target tissue in contact with the bloodstream bear **specific receptor proteins** for a particular hormone. Therefore, when a hormone reaches its target tissue, it evokes a specific response while other tissues lacking the receptor molecules remain unaffected.

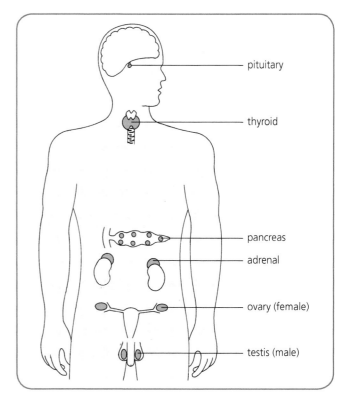

Figure 8.12 Endocrine glands

Blood glucose regulation

The body employs the control mechanism described below to guarantee that a regular supply of blood glucose is available for use by cells, regardless of when and how often food is eaten.

Liver as a storehouse

About 100 grams of glucose are stored as **glycogen** in the liver. Glucose can be added to or removed from this store depending on shifts of supply and demand.

Insulin and glucagon

A rise in blood glucose level to above its optimum concentration after a meal is detected by cells in the **pancreas**, which produce **insulin**. This hormone is transported in the bloodstream to the **liver** where it activates an enzyme that catalyses the reaction:

glucose $\longrightarrow$ glycogen

This brings the blood glucose concentration back down to around its normal level. If the blood glucose level drops below its optimum concentration between meals or during the night, different cells in the pancreas detect this change and release **glucagon**. This second hormone is transported to the liver and activates a different enzyme, which catalyses the reaction:

glycogen $\longrightarrow$ glucose

The blood glucose concentration therefore rises to around the optimum concentration. Figure 8.13 gives a summary of these events.

Osmotic imbalance

Control of blood glucose level is important in relation to **osmosis** in cells. The presence of too much or too little glucose in cells could lead to **osmotic imbalances**. For example, it could result in the cells gaining too much water from, or losing too much water to, the surrounding intercellular fluid by osmosis.

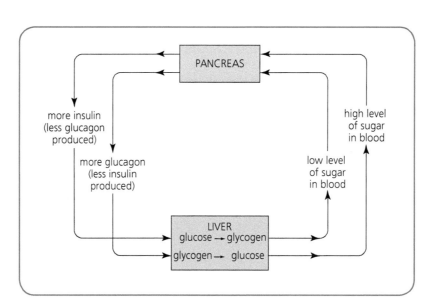

Figure 8.13 Regulation of blood glucose

Related Topic

Diabetes

People who suffer from **diabetes** are unable to control their blood glucose level. Diabetes is the result of a communication pathway that has broken down for one of the following reasons:

- the person's pancreatic cells are unable to make insulin (for example in people with type 1 diabetes)
- the target tissues in the person's body do not respond to insulin arriving in the bloodstream (for example in people with type 2 diabetes).

Consequences

If left untreated, diabetes results in a rapid **increase in blood glucose** concentration occurring after each meal. As the kidneys filter the blood, the filtrate formed is so rich in glucose that much of the glucose is excreted in **urine**. This increase in glucose concentration in urine is accompanied by the loss of additional water. The untreated diabetic produces excessive volumes of urine and is persistently thirsty. Over the long-term, people with **untreated** diabetes run a high risk of developing a variety of health problems including kidney disease and a form of eye condition that can result in blindness.

Treatment

Treatment for those people who fail to produce insulin takes the form of regular injections of insulin and a careful diet. Treatment for those people whose target tissues fail to respond to insulin takes the form of exercise, weight loss, diet control and additional insulin in some cases.

Causes

Type 1 diabetes is thought to be caused by a combination of some or all of the following:

- genetics
- environmental factors
- autoimmune factors.

Type 2 diabetes is thought to be caused by a combination of some or all of the following:

- unhealthy diet leading to obesity
- physical inactivity
- genetics.

To a certain extent, type 2 diabetes is a 'self-inflicted disorder'. Many cases could be prevented by people adopting a healthy lifestyle early in life and avoiding obesity throughout life.

Related Activity

Investigating trends in diabetic statistics

The number of people in Scotland with diabetes is increasing year on year, as shown in Figure 8.14.

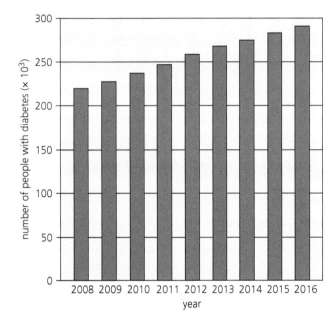

Figure 8.14 Incidence of diabetes

The results of a recent Scottish diabetes survey (2016) are shown in Table 8.1. It shows that:

- many more cases of type 1 than type 2 diabetes are found among children, teenagers and young adults
- many more cases of type 2 than type 1 diabetes are found to occur among middle-aged and older people
- overall, people with type 2 diabetes greatly outnumber those with type 1 diabetes.

Age (years)	Number of people registered with diabetes	
	type 1	type 2
0–9	749	0
10–19	3102	63
20–29	4890	963
30–39	4932	5267
40–49	5783	19573
50–59	5709	49759
60–69	3531	73178
70–79	1622	68606
80 and over	554	40218
totals	**30872**	**257627**

Table 8.1 Types of diabetes and age groups

Testing Your Knowledge 2

1 Rewrite the following sentences selecting only the correct answer from each underlined choice. (14)

a) Environmental stimuli are detected by <u>receptors</u>/<u>effectors</u>.

b) Responses to nerve impulses are brought about by <u>receptors</u>/<u>effectors</u>.

c) The first nerve cell in a reflex arc is called a <u>motor</u>/<u>sensory</u> neuron.

d) The final nerve cell in a reflex arc is called a <u>motor</u>/<u>sensory</u> neuron.

e) An electrical impulse is transmitted towards the cell body of a neuron by <u>an axon</u>/<u>a sensory fibre</u>.

f) An electrical impulse is carried away from the cell body of a neuron by <u>an axon</u>/<u>a sensory fibre</u>.

g) The gap between two neurons is called <u>a synapse</u>/<u>an inter neuron</u>.

h) A response made by a skeletal muscle is normally <u>faster</u>/<u>slower</u> than that made by a gland.

i) A rapid, automatic, involuntary response to a stimulus is called a reflex <u>arc</u>/<u>action</u>.

j) Chemical messengers secreted directly into the bloodstream are called <u>enzymes</u>/<u>hormones</u>.

k) Cells on a target tissue bear specific <u>receptors</u>/<u>effectors</u> for a particular hormone.

l) Insulin and glucagon are made in the <u>liver</u>/<u>pancreas</u>.

m) The conversion of glucose to glycogen in the liver is promoted by <u>glucagon</u>/<u>insulin</u>.

n) The conversion of glycogen to glucose in the liver is promoted by <u>glucagon</u>/<u>insulin</u>.

What You Should Know Chapter 8

action	diabetes	motor
arc	electrical	nervous
brain	endocrine	neurons
cerebellum	glucagon	pancreatic
cerebrum	glucose	receptor
chemicals	glycogen	sensory
CNS	hormones	slower
co-ordinated	insulin	synapses
cord	medulla	target

1 To survive, a multicellular animal needs the activities of all of its working parts to be _____. The _____ system provides the means by which the necessary internal communication is achieved.

2 In humans, the central nervous system (CNS) consists of the spinal _____ and the _____.

3 The brain is made up of the _____, which controls mental processes, the _____, which controls muscular co-ordination and the _____, which controls rate of breathing and heart beat.

4 The nervous system is composed of nerve cells called _____. A reflex _____ is an arrangement of a

sensory neuron, an inter neuron and a _____ neuron along which an _____ impulse passes from a _____ to an effector. A reflex _____ is a rapid, automatic response to a stimulus.

5 _____ are gaps between neurons where _____ are released allowing the transfer of an impulse from neuron to neuron.

6 _____ neurons pass information to the _____ where it is processed. Motor neurons transmit impulses to effectors enabling a rapid response to be made by a muscle or a _____ response to be made by a gland.

7 Chemical messengers released directly into the bloodstream by _____ glands are called _____. A hormone recognises its _____ tissue by the presence of specific receptors present on its cells.

8 The concentration of glucose in the bloodstream is regulated by two _____ hormones. Insulin promotes the conversion of glucose to _____ in the liver; _____ promotes the conversion of glycogen to _____ as required.

9 _____ is a disorder caused by the person's lack of ability to make _____ or the failure of their body cells to respond to it.

9 Reproduction

In multicellular animals and plants, normal body cells are **diploid** and sex cells (gametes) are **haploid** (see page 51).

Sexual reproduction

Reproduction is the process by which the members of a species produce offspring. **Sexual** reproduction involves the fusion of the nuclei of two **gametes** during **fertilisation**.

Gamete production in mammals

The human reproductive organs are shown in Figures 9.1, 9.2, 9.3 and 9.4. The **testes** are the site of **sperm** production; the **ovaries** are the site of **egg** production. Such sites of gamete production are called gonads.

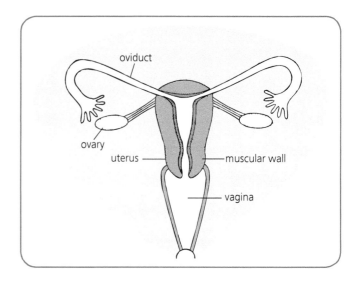

Figure 9.2 Human female reproductive organs

A male mammal produces a very large number of sperm and a female mammal produces a much smaller number of eggs. A sperm (see Figure 9.5) consists of

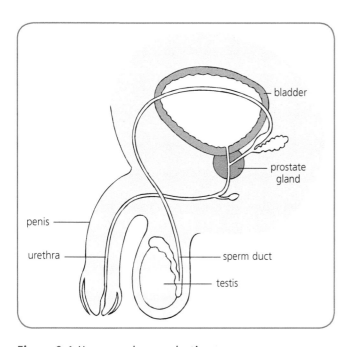

Figure 9.1 Human male reproductive organs

Figure 9.3 Sperm production in a human testis

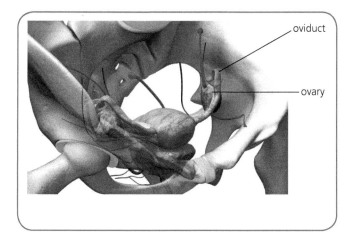

Figure 9.4 Model showing human ovaries and oviducts

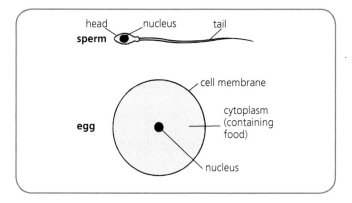

Figure 9.5 Sex cells (not drawn to scale)

a **head** region (mainly a nucleus containing genetic material) and a **tail** that enables it to move. An egg is larger than a sperm because in addition to its nucleus, it contains a **store of food** in its cytoplasm.

Events leading to fertilisation

The release of an egg from a mammalian ovary is called **ovulation**. Eggs are released at regular intervals into the oviducts (see Figure 9.6). The inner lining of an oviduct bears hair-like cilia that beat and gently move the egg along towards the uterus. The journey takes about three days in humans. Following copulation (sexual intercourse in humans) and ejaculation (release of sperm by the male), sperm swim up the uterus and into the oviducts. It is here in an **oviduct** that **fertilisation** normally occurs.

Fertilisation

This process involves a haploid sperm reaching a haploid egg and then the sperm's nucleus entering the egg and fusing with the egg's nucleus to form a single diploid cell called a **zygote.** The zygote formed contains genetic information from both parents and it is the first cell of a new individual. A zygote divides repeatedly by mitosis and cell division to become an **embryo.** (Note: although many sperm may meet an egg, only one sperm fertilises the egg.)

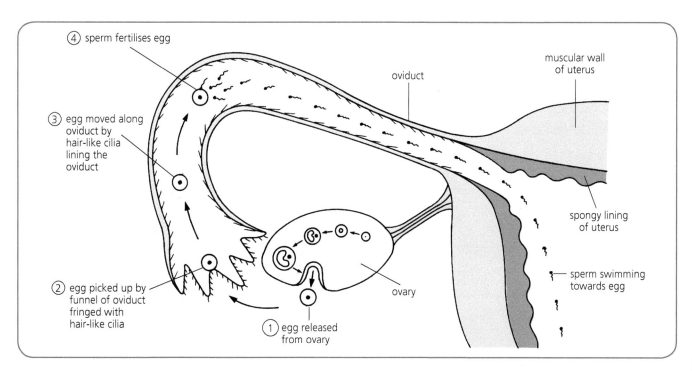

Figure 9.6 Events leading to fertilisation

Gamete formation in flowering plants

Flowers are the structures responsible for sexual reproduction in flowering plants. Although flowers often appear to be very different from one another (see Figures 9.7 and 9.8), they are built to the same basic plan. Usually the male and female reproductive organs are both present in the same flower (see Figure 9.9).

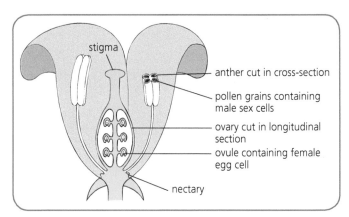

Figure 9.9 Half section of a flower

Pollen grains contain the plant's male sex cells and are produced in the **anthers**, the male sex organs of the flower. **Ovules** contain the plant's female sex cells (egg cells) and are produced in the **ovary**, the female sex organ of the flower.

Pollination

Pollination is the transfer of pollen grains from an anther to a stigma. It should not be confused with fertilisation, the process by which the nucleus of a male sex cell from a pollen grain fuses with the nucleus of an egg cell to form a zygote.

Growth of a pollen tube

Once a pollen grain has landed on a stigma, it responds to the presence of sugar on the stigma and uses the sugar as an energy source to form a **pollen tube** (see Figures 9.10 and 9.11). As the pollen tube grows down through the female tissues, the pollen grain's nucleus in the

Figure 9.7 Insect-pollinated flower

Figure 9.8 Wind-pollinated flower

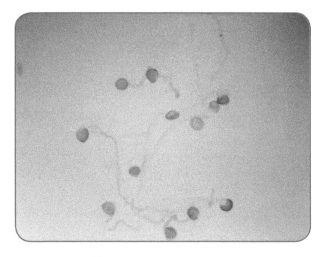

Figure 9.10 Pollen tubes

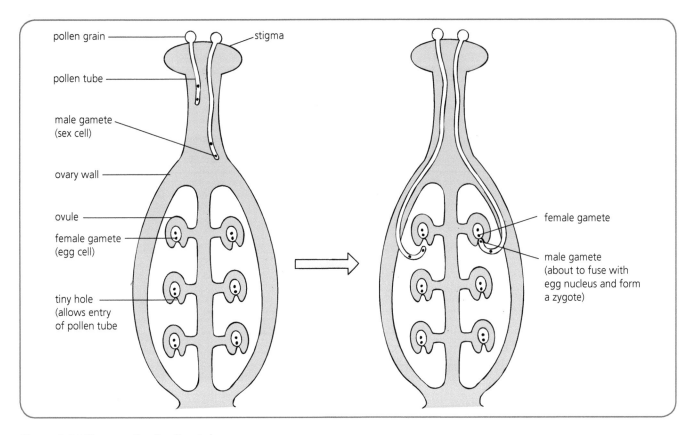

Figure 9.11 The growth of pollen tubes

region behind the tip of the tube divides into two male gametes. When the tip of the tube reaches an ovule, it enters through a tiny hole and comes into contact with a tiny sac containing the female gamete.

Fertilisation

The tip of the pollen tube bursts and the two haploid male gametes enter the sac. One fuses with the haploid egg cell to form a diploid **zygote.** (The other fuses with a second nucleus in the sac to form a type of cell that gives rise to the food store present in a seed.) Following mitosis and cell division, the zygote becomes an **embryo**, which grows into a plant when the seed containing it germinates.

Testing Your Knowledge

1 With reference to the relative numbers produced and their structure, draw up a table to show THREE differences between a mammalian sperm and an egg. (3)

2 a) Where are:
   i) sperm
   ii) eggs produced in the body of a mammal? (2)
   b) Identify the site of fertilisation in a mammal's body. (1)

3 Describe the process of fertilisation in a mammal using all of the following words in your answer: *diploid, egg, gamete, haploid, nucleus, sperm,* *zygote.* (You may wish to use some words more than once.) (4)

4 a) Which structures in a flower contain pollen grains? (1)
   b) Of which sex are the gametes present in the pollen grains? (1)
   c) Which structure in a flower contains egg cells in ovules? (1)
   d) By what means does a male sex cell reach and fertilise a female sex cell in a flower? (1)

10 Variation and inheritance

Although the members of the same species are very similar to one another, they are not identical. This is because **variation** exists among the members of a species.

Types of variation

Discrete variation

A characteristic shows **discrete variation** if it can be used to divide up the members of a species into two or more distinct groups. For example, ivy plants can be divided into two separate groups according to leaf type (green or variegated, as shown in Figure 10.1). Similarly, humans can be split into two separate groups depending on their ability or inability to roll their tongue (see Figure 10.2) and into four groups based on blood group types A, B, AB and O.

Continuous variation

A characteristic shows **continuous variation** when it varies among the members of a species in a **smooth, continuous way** from one extreme to the other and does not fall into distinct groups. Instead it shows a range of values between a minimum and a maximum. For example, seed length in broad beans varies continuously as shown in Figure 10.3. Similarly shell length in mussels (see Figure 10.4) and height in humans (see Figure 10.5) vary continuously.

Figure 10.2 Discrete variation in tongue-rolling ability

Figure 10.3 Continuous variation in broad bean length

Figure 10.1 Discrete variation in ivy leaf type

Figure 10.4 Continuous variation in mussel shell length

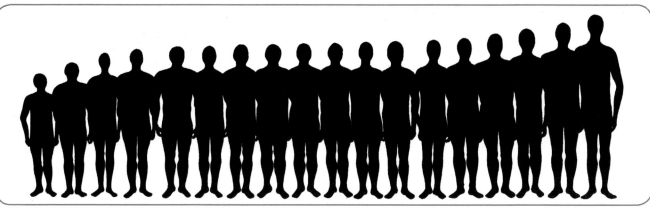

Figure 10.5 Continuous variation in male human height

Related Activity

Investigating a variety of discrete and continuous characteristics

Table 10.1 shows some further examples of variation.

Discrete variation	Continuous variation
• ear lobe type in humans (attached or unattached – see Figure 10.6) • fingerprint type in humans (loop, whorl, arch or compound – see Figure 10.7) • hair type in humans (straight, curly or wavy) • eye colour in fruit flies (red or white) • wing length in fruit flies (long or short) • coat colour in guinea pigs (black or white) • flower colour in foxglove plants (purple or white) • seed shape in garden pea plants (round or wrinkled) • height in garden pea plants (tall or dwarf) • colour of grain in maize plants (purple or yellow)	• body mass in humans • hand span in humans • foot length in humans • resting heart rate in humans • length of index finger in humans • height and diameter of shell in limpets (see Figure 10.8) • body length in trout • milk yield in cattle • mass of seeds in sunflower plants • mass of fruits on apple trees • length of petals in daisy flowers (see Figure 10.9) • breadth of tap root in dandelion plants

Table 10.1 Examples of discrete and continuous variation

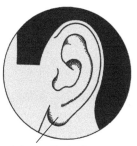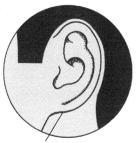

earlobe unattached   earlobe attached

loop   whorl

arch   compound

Figure 10.6 Discrete variation in ear type

Figure 10.7 Discrete variation in fingerprint type

Figure 10.8 Continuous variation in height and diameter of limpet shells

Figure 10.9 Continuous variation in length of petals in daisies

Presentation of data

Discrete variation

The data obtained from a survey of a characteristic that shows discrete variation is normally presented as a **bar graph**. Figure 10.10 shows a bar graph of the blood groups of 100 people with each distinct group represented by a separate bar.

Continuous variation

Table 10.2 shows the heights of the members of a sample group of 50 people arranged in increasing order. They do not fall naturally into distinct groups. For convenience, the entire range of the characteristic is divided into small groups (subsets) of 5 cm. For example, eight people are found to fall into the subset 155–159 cm whereas only two people fall into the subset 180–184 cm.

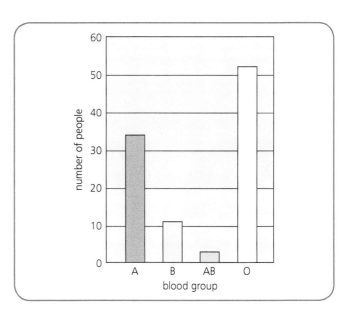

Figure 10.10 Bar graph of blood groups

Person	Height (cm)	Person	Height (cm)
1	144	26	164
2	145	27	164
3	148	28	164
4	151	29	165
5	152	30	165
6	152	31	165
7	153	32	166
8	154	33	166
9	155	34	167
10	155	35	167
11	156	36	168
12	156	37	168
13	157	38	169
14	158	39	170
15	158	40	171
16	159	41	172
17	160	42	172
18	160	43	174
19	161	44	174
20	161	45	176
21	162	46	178
22	162	47	179
23	162	48	181
24	163	49	183
25	163	50	186

Table 10.2 Continuous variation in human height

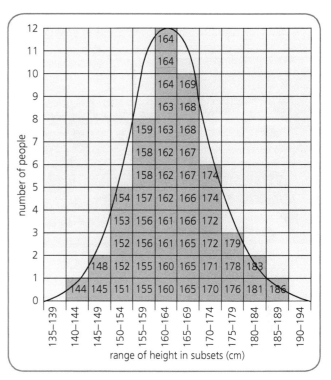

Figure 10.11 Histogram of human height

Variation and genetics

The genetic information responsible for the variation that exists among the members of a species is contained in genes on chromosomes (see Chapter 3). The study of this genetic material and its transmission from generation to generation is called **genetics**.

Significance of sexual reproduction

Each new member of a species formed by **sexual reproduction** receives half of its genetic material from one parent and half from the other parent. It therefore resembles each parent in some ways but differs from each parent in other ways. By combining genes from separate parents, sexual reproduction produces individuals that are genetically distinct from other members of the species and therefore it **contributes to variation within a species**.

Inheritance

Phenotype

Coat colour in guinea pigs, leaf shape in tomato plants, wing type in fruit flies and grain colour in maize plants, as shown in Figure 10.12, are a few examples of the many physical characteristics possessed by living

Grouping the data into subsets allows a **histogram** to be drawn, as shown in Figure 10.11. The majority of the people in the sample group have a height that is close to the centre of the range with fewer at the extremities. When a curve is drawn, a bell-shaped **normal distribution curve** is produced. In this example, the range in height extends from subset 140–144 cm to subset 185–189 cm and the most common height is the subset 160–164 cm.

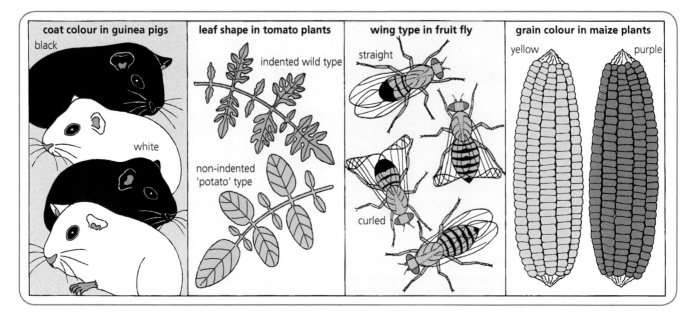

Figure 10.12 Phenotypes

organisms. These physical characteristics make up the organism's **phenotype**. Such characteristics are determined by genetic information.

True-breeding strains

Flower colour is a characteristic feature of pea plants that is controlled by genetic information. Purple and white are the two phenotypic expressions of flower colour in pea plants. Figure 10.13 shows three generations of a strain of pea plants with purple flowers; Figure 10.14 shows three generations of a strain of pea plants with white flowers. In each case the flower colour of the offspring is always identical to that of the parents and the members of the strain are described as being **true-breeding**.

Similarly, black guinea pigs that interbreed to form only black offspring and long-winged fruit flies that interbreed to produce only long-winged offspring are described as true-breeding strains.

Single gene inheritance

A characteristic such as coat colour in guinea pigs or grain colour in maize that shows **discrete variation** is controlled by different forms of a **single** gene. Geneticists often begin their investigation of inheritance in a species of plant or animal by studying a characteristic that shows discrete variation.

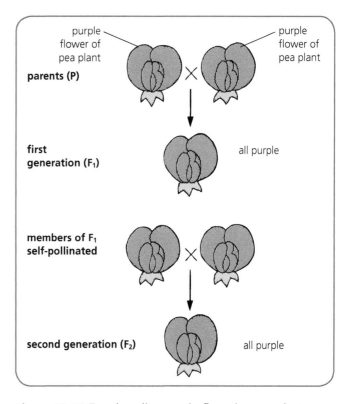

Figure 10.13 True-breeding purple-flowering pea plants

They set up a cross between two true-breeding strains of the organism that differ from one another in only one way. In guinea pigs, for example, a cross could be set up between two true-breeding strains differing only in coat colour. In fruit flies a cross could be set up between two true-breeding strains differing only in wing type. In pea

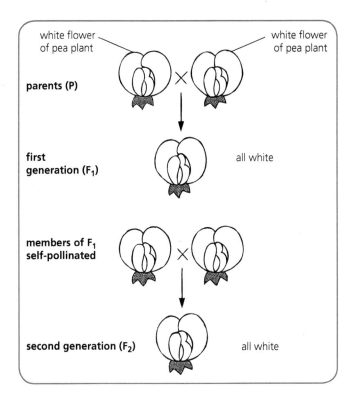

Figure 10.14 True-breeding white-flowering pea plants

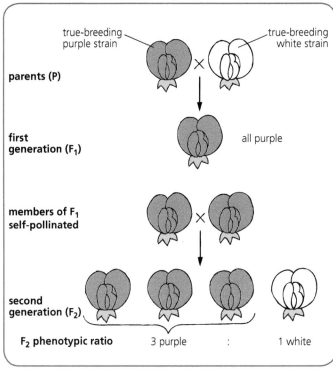

Figure 10.15 Single gene (monohybrid) cross in pea plants

plants, the example that follows – a cross between two true-breeding strains differing only in flower colour – is followed through two generations.

Inheritance of flower colour in pea plants

Figure 10.15 summarises a single gene (**monohybrid**) cross for flower colour in pea plants. The phenotypes of the members of the first (F₁) generation are always found to be the same. In this case, they all bear purple flowers.

The white characteristic has disappeared in the F₁ because it has been masked by the purple characteristic. Purple flower colour is therefore said to be the **dominant** characteristic and white colour the **recessive** one.

The second (F₂) generation does not contain any new 'in-between' forms of flower colour. Both purple and white flower colour appear in the F₂ generation in their original form, unaffected by their union in the F₁ generation. All single gene crosses of this type produce a **3:1 phenotypic ratio** in the F₂ generation.

Testing Your Knowledge 1

1 a) Explain what is meant by the terms *discrete* and *continuous* variation. (2)
  b) Give TWO examples of each type of variation that exist among a large population of humans. (4)
2 The crosses shown in Figure 10.16 refer to a genetics experiment using pea plants.
  a) Before being used in the experiment, the parent plants were tested to ensure that they were true-breeding. How would this be done? (2)
  b) Supply the symbols that would normally be used at positions X, Y and Z in Figure 10.16 to denote the different generations. (3)

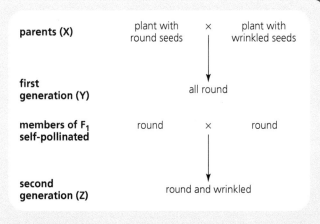

Figure 10.16

c) i) Which seed shape characteristic is recessive
and which is dominant?

ii) Explain how you arrived at your answer. (2)

d) What can always be said about the phenotypes of
the members of the F_1 generation resulting from a
cross between two true-breeding parents? (1)

e) Predict the ratio of the two phenotypes that would
occur in the F_2 generation. (1)

Genes and genotype

Each **gene** is a unit of heredity that controls an inherited
characteristic such as flower colour in pea plants (or
contributes in part to a polygenic characteristic such
as human height – see page 87). Each gene occupies a
specific site on a chromosome (see Figure 10.17). The
complete set of genes possessed by an organism is called
its **genotype**.

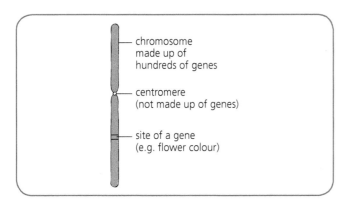

Figure 10.17 Site of a gene

Alleles

At least two forms of each gene normally exist among
the members of a species. For example, the gene
controlling flower colour in pea plants may be the
dominant form that produces purple flowers or the
recessive form that produces white flowers. These
different forms of a gene are called **alleles**.

Every normal body cell in an organism carries two
matching sets of chromosomes, one originating from
each parent. Thus every body cell has two alleles of
each gene, one from each parent. These two alleles
may be the same or different depending on which
alleles an organism inherits from its parents. Each
gamete, on the other hand, has only one set of
chromosomes and therefore carries only one allele of
each gene.

Symbols

For convenience, the dominant and recessive alleles of
a gene are often represented by **symbols**. In pea plants,
for example, the two alleles of the gene for flower colour
can be represented by the letters P for the dominant
allele (purple) and p for the recessive allele (white).
Since every **body cell** has two alleles of each gene, one
from each parent, an organism's genotype is always
represented by **two letters** per gene. A **gamete** only
carries one allele of each gene and is always represented
by **one letter** per gene. The cross in Figure 10.15 can be
represented as shown in Figure 10.18, which includes a
Punnett square.

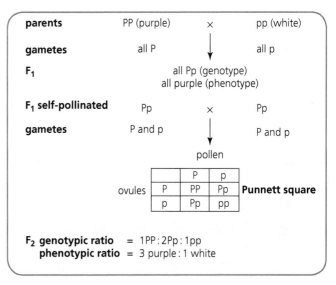

Figure 10.18 Use of symbols

Homozygous and heterozygous

When an organism possesses two identical alleles of a
gene (such as P and P or p and p), its genotype is said
to be **homozygous** and it is true-breeding. When an
organism has two different alleles of a gene (such as P
and p) its genotype is said to be **heterozygous** and it is
not true-breeding.

Same phenotype, different genotype

In the above example, if a pea plant's phenotype is white flowers then its genotype must be pp. However, if a pea plant's phenotype is purple flowers, its genotype may be PP or Pp. In other words, the two **different genotypes** PP and Pp have the **same phenotype** (purple).

Observed versus predicted figures for number of offspring

Single gene crosses of the type shown in Figure 10.18 always produce a 3:1 phenotypic ratio in the F_2 generation. However, there is often a difference between the **observed** and the **predicted** numbers of the different types of offspring, as shown in Figure 10.20 overleaf. In each of these single gene crosses, the F_2 offspring do not show the predicted ratio of 3:1 exactly, although each ratio is very close to it.

Continuing with flower colour in Figure 10.18 as our example, an exact 3:1 ratio would have occurred in the F_2 generation if during the Pp × Pp cross exactly half of the P ovules had been fertilised by P pollen grains and the other half of the P ovules by p pollen, while at the same time exactly half of the p ovules had been fertilised by P pollen and the other half by p pollen. However, this rarely happens in nature because fertilisation is a **random** process involving the element of **chance**.

| **Research Topic** | Mendel's work using pea plants |

Gregor Mendel (1822–84) carried out single gene crosses using true-breeding varieties of pea plant that possessed characteristics showing discrete variation (Figure 10.19). He carefully isolated plants and transferred pollen from one plant to another manually when cross-fertilisation was required. He collected and sowed the seeds formed by each generation of plants. He is regarded as being ahead of his time because he appreciated the importance of:

- working with **large numbers** of plants
- studying **one characteristic** at a time
- **counting the number** of different types of offspring produced and calculating a ratio.

By doing so, he was the first to put genetics on a firm scientific basis and his work is looked upon as the birth of genetics.

Figure 10.20 overleaf shows a summary of his single gene inheritance experiments. From the results of these experiments, Mendel drew the following conclusions (expressed in modern terms):

- The inheritance of characteristics is determined by factors (genes).
- In an individual organism, the alleles of a gene exist in pairs.
- At gamete formation, each gamete only receives one of a pair of alleles.
- Alleles retain their identity from generation to generation.

Figure 10.19 Seed shape (smooth or wrinkled) is one of the characteristics that Mendel studied

Characteristic	Dominant trait	Recessive trait	F₂ products	Ratio
flower colour	 purple	 white	705 purple, 224 white	3.15 : 1
seed colour	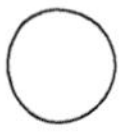 yellow	green	6022 yellow, 2001 green	3.01 : 1
seed shape	smooth	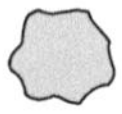 wrinkled	5474 smooth, 1850 wrinkled	2.96 : 1
pod colour	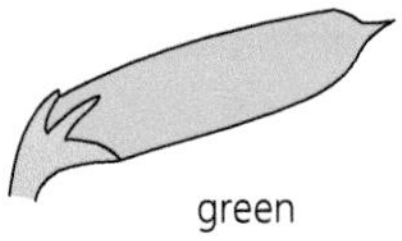 green	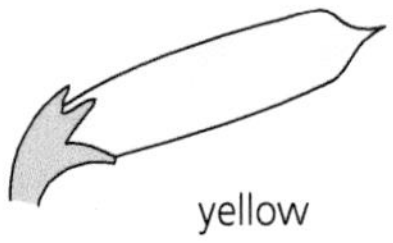 yellow	428 green, 152 yellow	2.82 : 1
pod shape	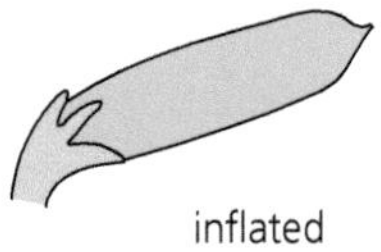 inflated	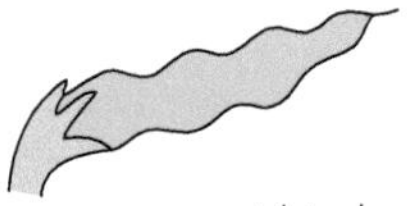 constricted	882 inflated, 299 constricted	2.95 : 1
flower position	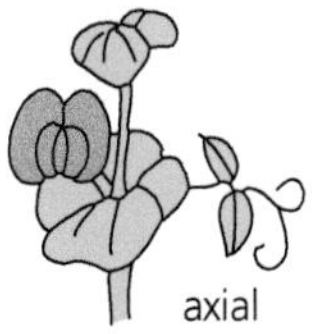 axial	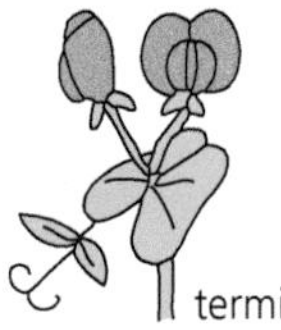 terminal	651 axial, 207 terminal	3.14 : 1
plant height	 tall	 dwarf	787 tall, 277 dwarf	2.84 : 1

Figure 10.20 Mendel's single gene crosses

Polygenic inheritance

A characteristic showing **continuous variation** is controlled by the alleles of more than one gene and is said to show **polygenic inheritance**. The more genes that are involved, the greater the number of intermediate phenotypes that can be produced.

Additive effect

The genes involved in polygenic inheritance are transmitted from generation to generation by sexual reproduction in the normal way. What makes them different from other genes is that their effects are **additive**. This means that each dominant allele of each gene **adds a contribution** towards the characteristic controlled by the genes.

Polygenic inheritance in humans

Many human characteristics such as height, weight, skin colour, hand span and foot size show a pattern of polygenic inheritance. Since there are many different expressions of height, weight, etc., each of these characteristics must be controlled by two or more genes. Detailed studies suggest that **three or more genes** are probably involved in each case, with every dominant allele making an individual contribution to the characteristic.

Environment

Whereas characteristics that show discrete variation (such as tongue rolling and ABO blood group) are unaffected by environmental factors, many characteristics that show continuous variation (such as height and foot size) are influenced by the **environment**. They are dependent on favourable environmental conditions for their full phenotypic expression. For example, regardless of how many dominant alleles for height a person inherits, they will not reach their full potential height without consuming an adequate diet during childhood and adolescence.

Thus a combination of polygenic inheritance and environmental factors produces phenotypic characteristics that show a wide range of continuous variation.

Family trees

Unlike pea plants, human beings do not breed to suit the geneticist. In addition, they produce too few offspring to allow reliable conclusions to be drawn about the phenotypic ratios produced. Nevertheless, the laws of genetics still apply to humans and particular traits can be traced through several generations of a family by constructing a **family tree** (pedigree chart).

Hair colour

In humans, the allele for red hair colour (h) is recessive to the dominant allele for non-red hair (H). If a person such as Sandy in Figure 10.21 has red hair yet neither of his parents has red hair, then it can be concluded that Sandy must have the **homozygous** genotype (hh) and that both of his parents must have the **heterozygous** genotype (Hh).

Since Sandy's aunt and maternal grandmother are red-haired they must be hh and his maternal grandfather must be Hh and so on. Piecing together such information for several generations of a family enables a geneticist to construct a family tree, as shown in Figure 10.21.

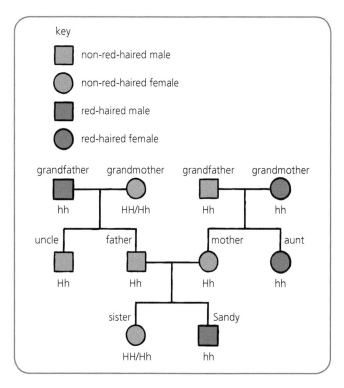

Figure 10.21 Family tree for hair colour

Element of chance

Each time Sandy's maternal grandparents produced a child, there was a 1 in 2 chance that the child would be non-red-haired (Hh) and a 1 in 2 chance that the child would be red-haired (hh).

Every time Sandy's parents (Hh and Hh) produce a child, there is a 1 in 4 chance that the child will be non-red-haired (HH), a 1 in 2 chance that the child will be non-red-haired (Hh) and a 1 in 4 chance that the child will be red-haired (hh).

Related Topic

Use of family trees

A family tree may be constructed and used by a **genetic counsellor** when advice is needed by a couple who are concerned about passing on a genetic disorder (known to exist in their families) to their children.

Cystic fibrosis

Cystic fibrosis is a genetic disorder caused by a change (mutation) to one gene. It results in the sufferer producing mucus that is thicker and stickier than normal. This causes congestion of the lungs and other organs.

Let us consider the case of a couple (Mary and Jack) who want advice about the risk of their children inheriting the allele of the gene for cystic fibrosis. A genetic counsellor would interview the couple and construct their family trees as shown in Figure 10.22. She would then be able to deduce the genotypes and add them to the tree as shown in Figure 10.23 (where M = normal allele and m = allele for cystic fibrosis).

She would conclude that both Mary and Jack are heterozygous for this gene and are therefore **carriers** of the recessive cystic fibrosis allele. The counsellor would then advise the couple that each child they produce would have a 1 in 4 chance of suffering cystic fibrosis and a 1 in 2 chance of being a carrier of the allele for the disorder.

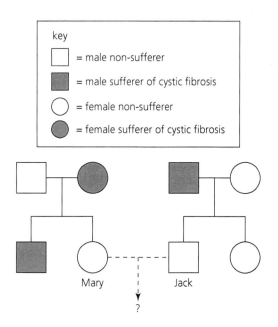

Figure 10.22 Family tree with phenotypes

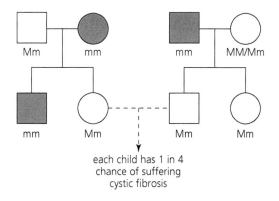

Figure 10.23 Family tree with deduced genotypes

Testing Your Knowledge 2

1 a) What name is given to the basic unit of inheritance that controls a characteristic that is passed on from generation to generation? (1)
  b) Of what structures found in a cell's nucleus do such units form a part? (1)

2 State the meaning of the terms *genotype* and *phenotype*. (2)

3 a) What term is used to refer to the different forms of a gene? (1)
  b) How many forms of a gene does a zygote receive from each parent? (1)

4 Explain the difference between the terms *homozygous genotype* and *heterozygous genotype* with reference to a named example. (2)

5 a) i) Which type of variation, discrete or continuous, is controlled by alleles of more than one gene?
     ii) What term is used to refer to this type of inheritance? (2)
  b) Apart from genetic factors, what other type of factors influences the wide range of phenotypic expression shown by some characteristics such as human body mass? (1)

What You Should Know Chapters 9–10

alleles	genes	phenotype
anthers	genotype	pollen
chromosome	haploid	polygenic
continuous	heterozygous	recessive
diploid	homozygous	smaller
discrete	more	sperm
dominant	numerous	testes
egg	ovaries	uninterrupted
fertilisation	ovules	zygote

1 In a multicellular organism, the body cells are _____ and the sex cells (gametes) are _____.

2 In male animals, the _____ are the site of _____ production; in females, the _____ are the site of egg production.

3 Sperm are _____ in size and more _____ than eggs.

4 In flowering plants, the male gametes are contained in _____ produced in _____. The female sex cells (_____ cells) are contained in _____ produced in the ovary.

5 _____ is the process by which a haploid male gamete fuses with a haploid female gamete to produce a diploid _____.

6 Variation exists among the members of a species. When a characteristic can be used to divide the species into distinct groups, it is said to show _____ variation. When the characteristic varies in an _____ way from one extreme to the other, it is said to show _____ variation.

7 An organism's physical characteristics are known collectively as its _____.

8 Each inherited characteristic is controlled by one or more units of heredity called _____. Each gene is part of a _____.

9 Each gene normally has two or more different forms called _____. An allele that always shows its effect and masks the presence of the other form is said to be _____. An allele that is masked by the dominant form is said to be _____.

10 The complete set of genes possessed by an organism is called its _____. The genotype of an organism with identical alleles of a gene is described as _____; the genotype of an organism with two different alleles of a gene is described as _____.

11 A characteristic that shows continuous variation is controlled by alleles of _____ than one gene and is said to show _____ inheritance.

11 Transport systems – plants

Plant transport systems

Green plants need water for transporting materials and for photosynthesis. A land plant's roots are buried in the ground where they absorb water and mineral salts. The green leaves are above ground where they make sugar by photosynthesis. Therefore the plant needs **two transport systems** (see Figure 11.1) to enable water to travel up to the leaf cells and sugar to pass down to the root cells.

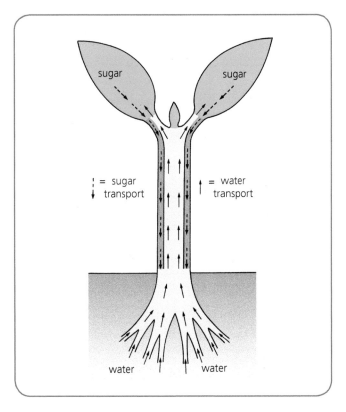

Figure 11.1 Plant transport systems

Related Activity

Investigating the site of water transport in a plant

Stem structure

Figure 11.2 shows where the **xylem** and **phloem** tissue are found in the stem of a plant such as *Impatiens* (Busy Lizzy) that is a few months old and in the woody stem of a plant such as privet that is one year old.

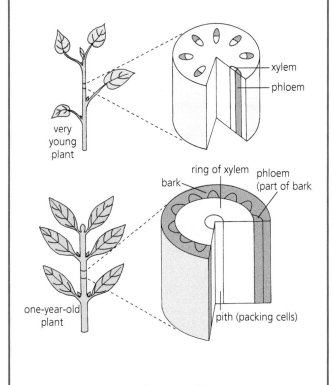

Figure 11.2 Sites of xylem and phloem in stems

Red dye experiment

The cut end of a leafy shoot from each type of plant is placed in **red dye** for at least 30 minutes. A portion of stem is then cut out and examined as shown in Figure 11.3. Red dye is found to be present only in the xylem cells (vessels) showing that **xylem** is the site of **water transport** in a plant. (A set of different experiments, not shown here, can be used to demonstrate that phloem is the site of sugar transport in a plant.)

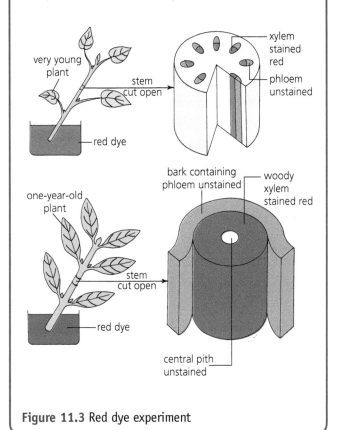

Figure 11.3 Red dye experiment

Water transport in a plant

Figure 11.5 (overleaf) shows the route taken by water as it passes up through a plant from root to leaf. Figure 11.4 shows **root hairs** on the young roots of germinating seeds. A root hair is an extension of an epidermal cell and it presents a **large absorbing surface** to the surrounding soil solution.

Since a root hair is a region of low water concentration compared with the soil solution (a region of higher water concentration), water enters a root hair by **osmosis**. Water then passes from the root hair (which now has a higher water concentration) into a neighbouring cell (which by comparison has a lower water concentration) and so on across the root to the **xylem** vessels. The xylem vessels are hollow, dead tubes. They transport **water** and **mineral salts** up through the stem to all parts of the plant. **Mesophyll** cells in the leaf need water for photosynthesis.

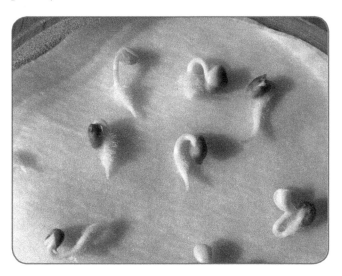

Figure 11.4 Root hairs

Exit of water

Transpiration is the process by which water is lost by evaporation from the aerial parts of a plant. Most transpiration occurs through tiny pores in the leaves called **stomata**.

The rate of transpiration is affected by factors such as the surface area of leaf exposed, temperature, air humidity and wind speed (see the Related Activity on page 93).

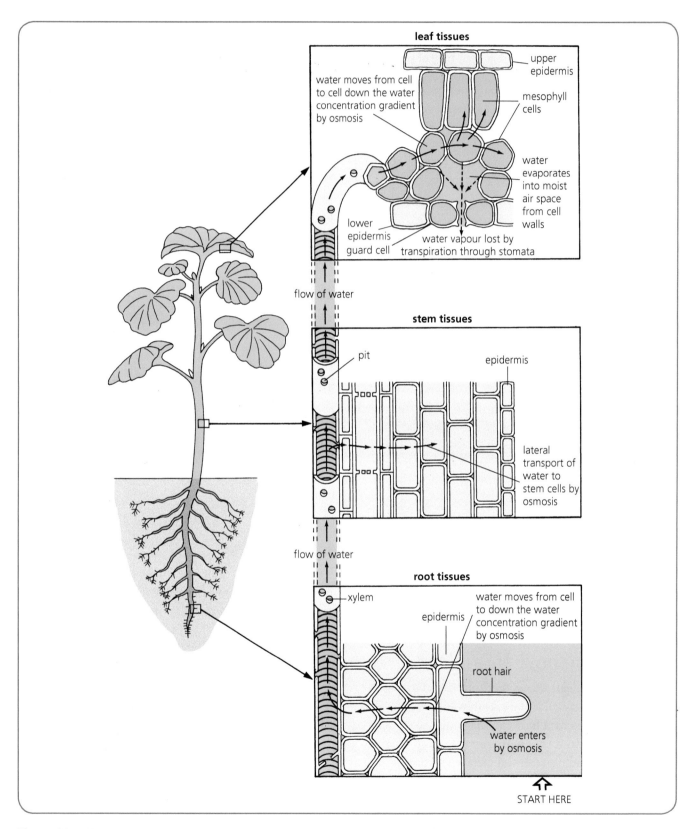

Figure 11.5 Water transport system

Stomatal mechanism

Guard cells (see Figure 11.8 overleaf) differ from other epidermal cells in three ways. Guard cells are **sausage-shaped** and possess **chloroplasts**. In addition, the inner regions of their cell walls (those facing the stomatal pore) are **thicker** and **less elastic** than the outer regions of their cell walls.

Opening and closing of stomata occur as a result of **changes in turgor** of guard cells. When water enters a pair of flaccid guard cells, turgor increases. Due to their larger surface area and greater elasticity, the thin outer parts of the two guard cells' walls become stretched more than the thick inner parts. As the two guard cells bulge out, the thick inner walls become pulled apart, **opening** the stoma. This occurs in **light**.

When water leaves a pair of guard cells, they lose turgor and return to a flaccid condition. This results in the **closing** of the stoma. It occurs in **darkness**.

Related Activity

Investigating transpiration

Site of transpiration

When the experiment shown in Figure 11.6 is examined after several days, much condensation is found to be present on the inner surface of bell jar A, very little on the inside of B and none in C. It is therefore concluded that most water is lost by a plant's **leaves** during transpiration.

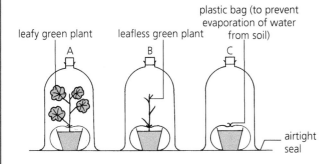

Figure 11.6 Investigating the site of transpiration

Rate of transpiration

A **bubble potometer** (see Figure 11.7) is an instrument used to measure **rate of water uptake** by a leafy shoot. This water uptake is only approximately equal to the transpiration rate because some water is normally retained by the leafy shoot for other processes such as photosynthesis.

An **air bubble** is allowed to enter the system and its rate of movement along the horizontal tube is measured (for example in mm/min). The syringe is used to inject water and return the bubble to the start of the scale, allowing the experiment to be repeated. Table 11.1 (overleaf) gives the reasons for adopting certain techniques and precautions during the investigation. Table 11.2 gives a specimen set of results for three different environmental conditions.

From the results it is concluded that **wind** increases the rate of transpiration. This occurs because the air outside the stomata is continuously being replaced with drier air that accepts more water vapour from the plant. Increased **humidity** of the air surrounding the plant results in decreased transpiration rate. This occurs because the difference in concentration of water vapour between the inside and the outside of the leaf is reduced and therefore rate of diffusion of water molecules from the plant slows down.

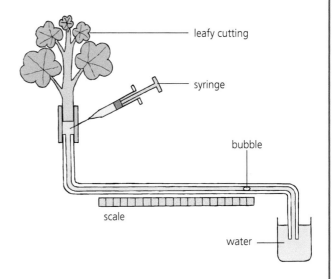

Figure 11.7 Bubble potometer

Design feature or precaution	Reason
stem cut under water and end of stem, rubber tubing and capillary tubing all connected under water	to prevent air entering xylem and forming air locks
tightly fitting rubber tubing used	to prevent leakage of water and ensure that system is airtight
time allowed for plant to equilibrate between different environmental conditions	to ensure that rate of movement of bubble is governed by factor being investigated and not the previous one
repeat measurements of rate of movement of bubble taken for each condition and average calculated	to obtain a more reliable result for each condition
all factors kept equal except for one change in environmental conditions	to ensure that the experiment is valid by only altering one variable factor at a time

Table 11.1 Experimental design features and precautions

Environmental condition	Additional apparatus needed to create condition	Average rate of movement of bubble (mm/min)
normal day	none	5
windy day	electric fan	20
humid day	transparent plastic bag	1

Table 11.2 Bubble potometer results

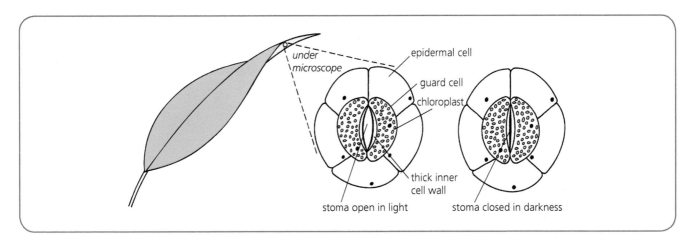

Figure 11.8 Stomata in light and dark

Related Activity

Investigating stomata

Location of stomata

When a leaf from a non-grass-like plant such as geranium is immersed in very hot water, many tiny bubbles appear on the leaf's lower surface and a very few on its upper surface. Each bubble indicates the **position of a stoma** through which hot expanded air has escaped from a **moist air space** (see Figures 11.4 and 11.12). It is concluded that in this type of plant, most stomata are located on the **underside** of the leaf.

When the procedure is repeated using a leaf from a grass, approximately equal numbers of tiny bubbles appear on both surfaces. It is therefore concluded that the stomata are **equally distributed** on the two surfaces of a grass leaf.

Structure of a stoma

A **stoma** is a slit-like pore enclosed by two **guard cells** (see Figure 11.9). Stomata and their guard cells can be seen under a microscope by preparing and viewing a **leaf peel** (see Figure 11.10). Figure 11.11 shows a microscopic view of a leaf surface using a thin layer of nail varnish.

Model of a stoma

When a sausage-shaped balloon that has had a strip of adhesive tape stuck to one of its sides is inflated, it bends over towards the reinforced side. Two balloons treated in this way and tied together at their ends act as a model of an open stoma.

Figure 11.9 Stoma and guard cells

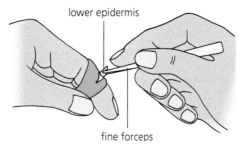

lower epidermis

fine forceps

Figure 11.10 Preparing an epidermal peel from a leaf

Figure 11.11 Leaf surface as revealed by a layer of nail varnish

Leaf structure

Figure 11.12 gives a summary of the **internal structure** of a leaf and the functions of its various parts.

Sugar transport

Sugar is transported through a plant in **phloem** cells. This occurs in both an **upward** and a **downward** direction depending on circumstances. For example, sugar made in leaves by photosynthesis may be transported up to the plant's growing points and flowers or down to its stem and roots. Furthermore, starch in a storage organ such as a bulb may be converted to sugar and transported up to growing leaves and flowers.

Comparison of xylem and phloem

Figure 11.14 (on page 97) shows both types of transport tissue in longitudinal section. **Xylem** is composed of non-living, hollow tubes (vessels) that are described as being **lignified** because they bear rings or spiral bands of strong **lignin** (see Figure 11.13 overleaf). These structures provide **support** and enable the xylem vessels to **withstand pressure changes** that occur when water is drawn up through them to replace water lost by transpiration.

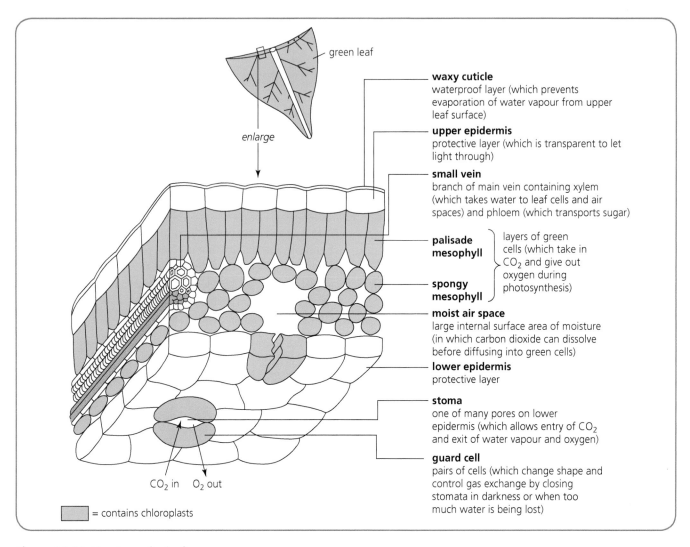

Figure 11.12 Structure of a leaf

Phloem is living tissue composed of **sieve tubes** and **companion cells**. The cytoplasm in sieve tubes is continuous from cell to cell through holes in sieve plates (the name given to the perforated walls at the ends of the sieve tubes). The continuity of cytoplasm allows sugar to be transported to all parts of the plant. Sieve tubes lack nuclei and their activities are controlled by neighbouring companion cells.

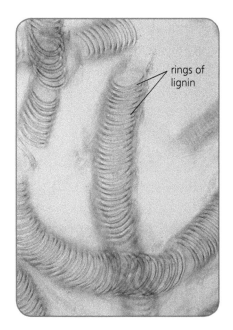

Figure 11.13 Xylem

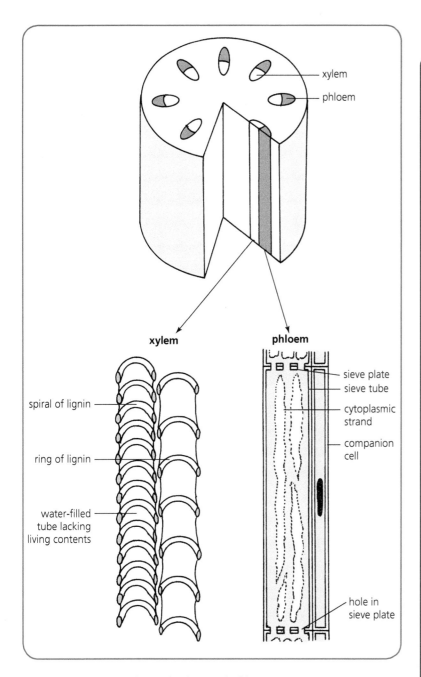

Figure 11.14 Comparison of xylem and phloem

Labels in figure:
- xylem
- phloem
- xylem
- phloem
- spiral of lignin
- ring of lignin
- water-filled tube lacking living contents
- sieve plate
- sieve tube
- cytoplasmic strand
- companion cell
- hole in sieve plate

Testing Your Knowledge

1 Explain why a plant needs two different transport systems. (2)
2 Copy and complete Table 11.3. (6)

Type of cell or tissue	Description
	absorbs water from soil solution
	lignified to withstand pressure changes during transport of water
	main site of photosynthesis in a leaf
	controlled by a companion cell
	protective layer on outside of plant organs
	controls opening and closing of stomata

Table 11.3

3 Which ONE of the following is correct? (1)
Transpiration is the:
A transport of water up the xylem vessels
B movement of sugars up to new growing points
C transport of glucose down sieve tubes
D evaporation of water from leaf surfaces.
4 Name TWO environmental factors that can affect the rate of transpiration. (2)
5 a) Construct a table to compare the transport of substances in phloem and xylem with respect to:
  i) type of substance transported
  ii) direction taken by substance
  iii) state of tissue (dead or alive). (6)
 b) Name ONE other function of xylem in addition to transport. (1)

12 Transport systems – animals

Once essential substances such as oxygen and glucose have entered the animal's body, they must be carried to all of its living cells at a rate faster than is possible by diffusion. In mammals, this rapid transport of essential materials is achieved by the animal's **circulatory system**.

Mammalian circulatory system

The circulatory system consists of the **heart** (a muscular pump) and the **blood vessels** (a system of tubes), which carry **blood** to all parts of the body. Nutrients, oxygen, carbon dioxide and hormones are transported in the blood.

Heart

The **heart** is divided into two separate sides, as shown in Figure 12.1. Each side has two hollow chambers: an **atrium** (plural **atria**) and a **ventricle**. The diagram shows the four heart chambers viewed from the front of the person. The right-hand side of the heart is therefore on the left side of the diagram and vice versa. The wall of the heart is made of cardiac muscle.

Blood flow

The path taken by blood as it flows through the heart and its associated blood vessels is shown in Figure 12.2. Blood from the body enters the right atrium by the main vein called the **vena cava**. The right atrium contracts, pumping blood into the right ventricle. Contraction of the muscular ventricle wall forces blood into the **pulmonary artery**. This vessel carries blood from the heart to the lungs where it takes up oxygen.

The **pulmonary vein** returns oxygenated blood from the lungs to the heart's left atrium. The blood is then pumped into the left ventricle where strong contraction of the muscular wall forces blood into the main artery called the **aorta**. The aorta takes blood to all parts of the body. These tissues and organs remove the oxygen as the blood passes through, making it become deoxygenated. Deoxygenated blood returns by the vena cava to the heart's right atrium to repeat the cycle.

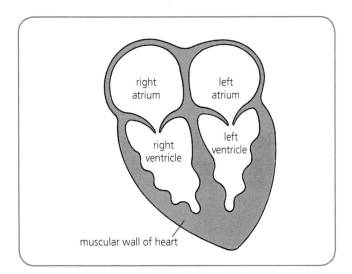

Figure 12.1 Chambers of the heart

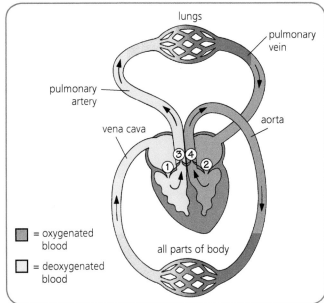

Figure 12.2 Heart and associated blood vessels (1–4 show the positions of heart valves)

Related Activity

Investigating heart structure

With the hands protected by disposable gloves, a **sheep's heart** is held with both hands to examine it closely, feel its muscular wall and observe its associated blood vessels. The heart is then returned to the teacher for dissection.

Figure 12.3 shows a sheep's heart before dissection and after the front half has been cut away. The heart's four chambers are numbered 1–4 in the diagram.

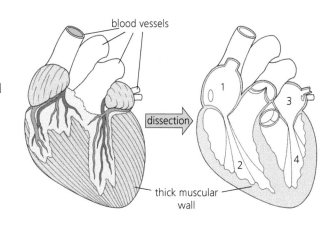

Figure 12.3 Structure of a sheep's heart

Thickness of ventricle walls

The wall of the left ventricle is particularly thick and muscular since it is required to pump blood all round the body. The wall of the right ventricle is less thick since it only pumps blood to the lungs.

Valves

Figure 12.2 shows the positions of the four heart **valves**. Valves 1 and 2 are situated between the atria and the ventricles. When they are open, blood passes from the atria into the ventricles. When the ventricles contract, the blood, under pressure, closes valves 1 and 2. This prevents blood from flowing back into the atria.

Valves 3 and 4 are situated between the ventricles and the two arteries that leave the heart. Once blood has been pumped through valves 3 and 4 they close, **preventing backflow** of blood from the arteries into the ventricles. Blood is therefore only able to travel in **one direction** through the heart. Figure 12.4 shows an artificial heart valve.

Blood vessels

Arteries and veins

An **artery** is a vessel that carries blood **away** from the heart. It has a **thick muscular wall** (see Figure 12.5 overleaf) that is able to withstand the **high pressure**

Figure 12.4 Artificial heart valve

of oxygenated blood coming from the heart. (The pulmonary artery is exceptional in carrying deoxygenated blood.) Each time the heart beats, blood is forced along the arteries at high pressure and this pressure wave can be felt as a **pulse** beat.

A **vein** is a vessel that carries blood **towards** the heart. Although muscular, its wall is thinner than that of an artery (see Figure 12.5) since deoxygenated blood

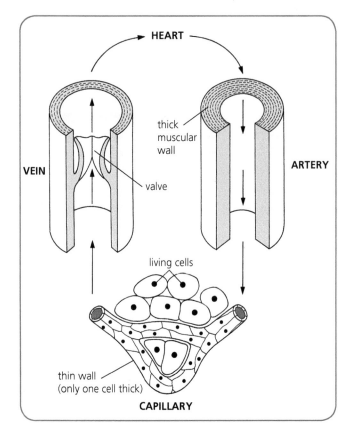

Figure 12.5 Types of blood vessel

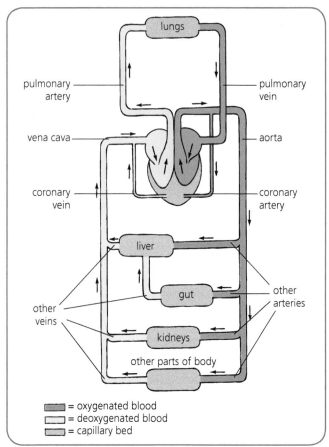

Figure 12.6 Human circulatory system

flows along a vein at **low pressure**. (The pulmonary vein is exceptional in carrying oxygenated blood.) Compared with an artery, the central cavity of a vein is wider. This helps to reduce resistance to the flow of blood along a vein. **Valves** are present in veins to prevent backflow.

Capillaries

An artery divides into smaller vessels and finally into a dense network of tiny, thin-walled **capillaries** (see Figure 12.5). Capillaries are the most numerous type of blood vessel in the body. They present a **large surface area** and are in close contact with the living cells in tissues and organs.

Capillaries are often referred to as **exchange vessels** since all exchanges of materials between blood and living tissues take place efficiently through their **thin walls** (only one cell thick). Capillaries unite to form larger vessels that converge to form veins. A simplified version of the human circulatory system is shown in Figure 12.6.

Coronary artery

The first branch of the main artery (aorta) leaving the heart is called the **coronary artery** (see Figures 12.6, 12.7 and 12.8). This artery supplies the muscular wall of the heart itself with oxygenated blood. If this vessel becomes blocked, blood flow to the heart wall is prevented and the person suffers a **heart attack**.

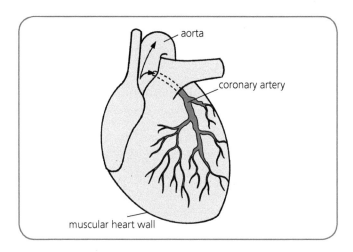

Figure 12.7 Coronary artery

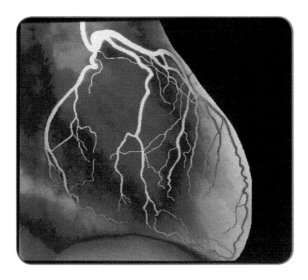

Figure 12.8 Details of the coronary artery

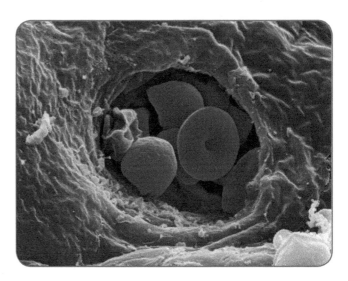

Figure 12.9 Red blood cells in a capillary

Blood

Mammalian blood contains plasma, red blood cells and white blood cells.

Plasma

Plasma is a watery yellow fluid that contains many dissolved substances, such as glucose, amino acids and respiratory gases.

Red blood cells

Red blood cells (see Figure 12.9) are very small and extremely numerous (approximately 5 million per mm^3). They are specialised in the following ways. They do not contain a nucleus. Their biconcave disc shape (see Figure 12.10) presents a **large surface area** for the uptake of oxygen. They contain red pigment called **haemoglobin**. As blood passes through lung capillaries,

haemoglobin (dark red in colour) combines with oxygen to form **oxyhaemoglobin** (bright red in colour). When blood reaches the capillaries beside respiring cells, oxyhaemoglobin quickly releases the oxygen, which then diffuses into the cells (see Figure 12.11).

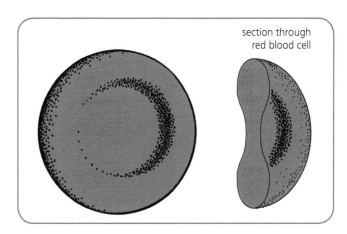

section through
red blood cell

Figure 12.10 Shape of a red blood cell

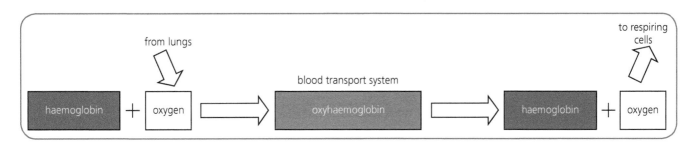

from lungs

blood transport system

to respiring cells

haemoglobin + oxygen → oxyhaemoglobin → haemoglobin + oxygen

Figure 12.11 Role of haemoglobin

White blood cells

White blood cells are part of the human body's immune system. They bring about the destruction of **pathogens**. Pathogens are micro-organisms such as bacteria, fungi and viruses that can cause disease if they manage to enter the body. The two main types of white blood cell involved in the destruction of pathogens are **phagocytes** and **lymphocytes**.

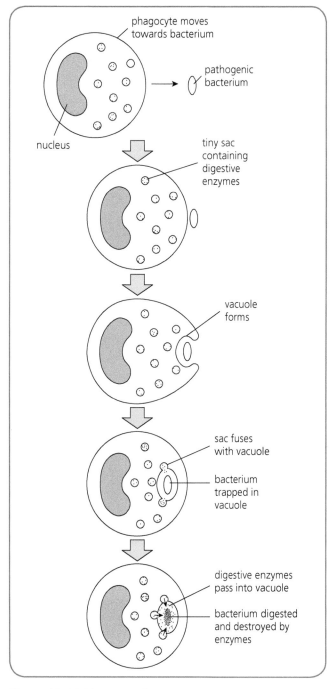

Figure 12.12 Phagocytosis

Phagocytes

Phagocytes engulf and destroy pathogens by **phagocytosis** as shown in Figure 12.12. During infection, many phagocytes migrate to the infected area and engulf the pathogens. Dead bacteria and phagocytes often accumulate at the site of infection, forming **pus**.

Lymphocytes

Some lymphocytes produce antibodies in response to the presence of a pathogen bearing antigens on its surface. An **antigen** is a molecule such as a protein that is recognised by the human body as foreign. An **antibody** is a Y-shaped protein molecule (see Figure 12.13). Each of an antibody's arms bears a receptor (binding site) that is **specific** to (exactly fits) a particular antigen.

The action of antibodies is shown in Figure 12.14. The binding of antibodies to a pathogen's antigens inactivates the pathogen in preparation for its destruction by phagocytosis.

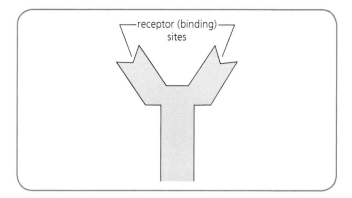

Figure 12.13 Antibody

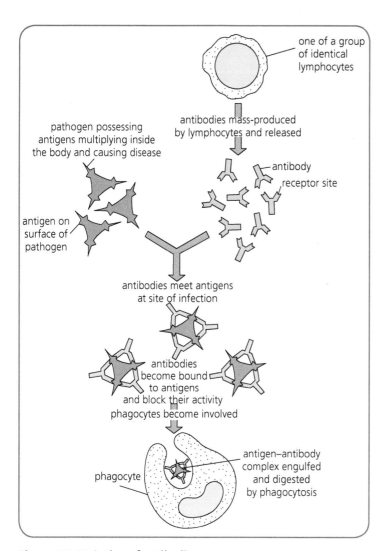

one of a group
of identical
lymphocytes

antibodies mass-produced
by lymphocytes and released

pathogen possessing
antigens multiplying inside
the body and causing disease

antibody

receptor site

antigen on
surface of
pathogen

antibodies meet antigens
at site of infection

antibodies
become bound
to antigens
and block their activity
phagocytes become involved

phagocyte

antigen–antibody
complex engulfed
and digested
by phagocytosis

Figure 12.14 Action of antibodies

Testing Your Knowledge

1 a) Name the FOUR chambers of the human heart. (2)
   b) Which of the simplified diagrams of a mammalian heart
      in Figure 12.15 shows the blood flowing in the correct
      direction? (1)

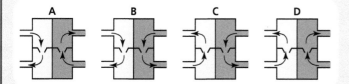

A   B   C   D

Figure 12.15

   c) i) Identify blood vessels 1, 2 and 3 in Figure 12.16.
      ii) From where has the blood in vessel 1 come?
      iii) Why is a blood clot at position A or B of particular
           danger to the affected person? (5)

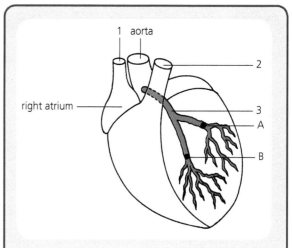

1 aorta

2

right atrium

3
A

B

Figure 12.16

2 Draw a table to show THREE structural
   differences between an artery and a vein. (3)
3 a) Name the type of blood vessel shown in
      Figure 12.17. (1)
   b) i) Identify the type of blood cell
          present in the vessel.
      ii) What is the red pigment that this
          type of cell contains?
      iii) Briefly describe the function of the
           red pigment. (4)

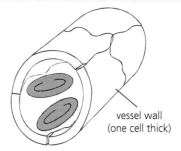

vessel wall
(one cell thick)

Figure 12.17

4 Describe how a phagocyte carries out its
   function. (3)
5 Which type of blood cell is able to make
   antibodies to defend the body? (1)

103

13 Absorption of materials

A large multicellular animal has a small surface area relative to its volume. Therefore, it needs **additional absorbing areas** to allow an adequate uptake of oxygen and food. In human beings, the internal structure of the **lungs** and of the **small intestine** greatly increase the surface area available for the absorption of oxygen and dissolved food, respectively. On entering the bloodstream, these essential substances are transported to cells for use in respiration. **Capillary networks** in tissues provide the large surface area needed for the absorption of food and oxygen by respiring cells, and the release of wastes such as carbon dioxide (see Figure 13.1).

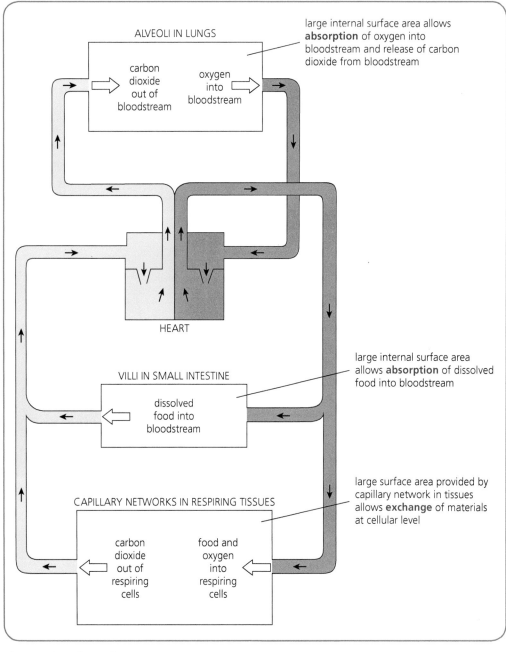

Figure 13.1 Absorption

Organs of gaseous exchange

The **lungs** (see Figure 13.2) are a mammal's organs of gaseous exchange.

Internal structure of the lung

Air passing along the narrow bronchioles ends up in tiny air sacs deep in the lungs (see Figure 13.3). These air sacs are called **alveoli** (singular alveolus). The alveoli are so numerous that they provide a very **large surface area** for gas exchange. The total internal surface area of the two lungs is approximately 90 m² (roughly the area of one side of a tennis court).

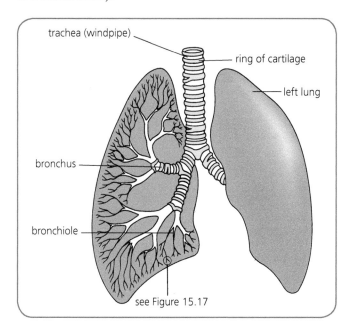

Figure 13.2 Organs of gas exchange

Capillary network

Each alveolus is surrounded by a **dense network** of blood capillaries. The lining of an alveolus is very **thin** and in **close proximity** to the walls of the blood capillaries, which are themselves only one cell thick. This combination of:

- large surface area
- short distance
- thin walls
- good blood supply

presents ideal conditions for **gas exchange** to occur between alveolar air and blood.

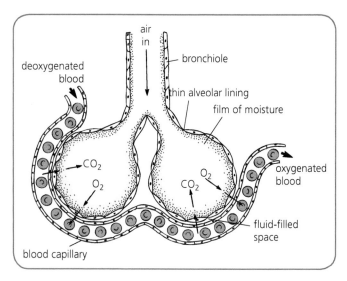

Figure 13.3 Gas exchange in alveoli

Gas exchange

Blood arriving in a lung is described as **deoxygenated** because it contains a low concentration of oxygen. Since air breathed into an alveolus contains a higher concentration of oxygen, diffusion occurs. Oxygen first dissolves in the moisture on the inner surface of the thin lining of an alveolus (see Figure 13.3) and then diffuses into the blood in the surrounding capillaries. The blood therefore becomes **oxygenated** (rich in oxygen) before leaving the lungs and passing to all parts of the body.

Deoxygenated blood contains a higher concentration of carbon dioxide than the air entering the alveoli. Therefore carbon dioxide diffuses from the blood into the alveoli from where it is exhaled.

Food transport system

The alimentary canal (see Figure 13.4 overleaf) is the body's **digestive system**. As insoluble molecules of food pass along this muscular tube they are broken down to a soluble state by digestive enzymes.

End products of digestion

Imagine the small intestine of a person who has eaten an egg sandwich. As a result of complete digestion, the starch in the bread has been broken down to **glucose**, the protein in the egg to **amino acids** and the fat in the

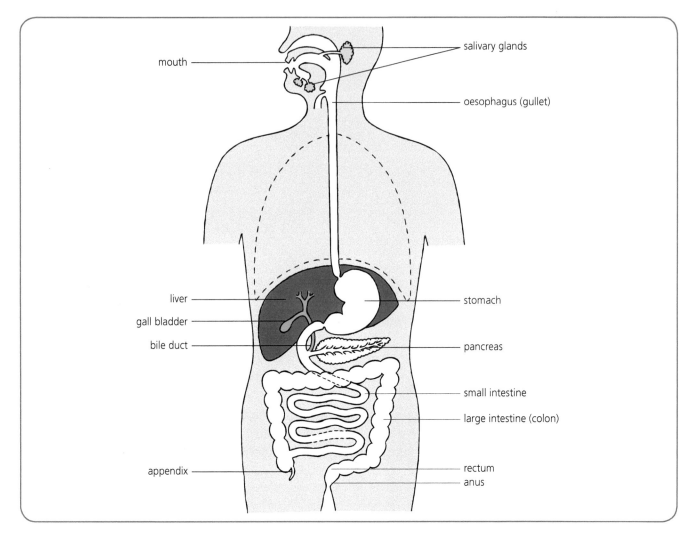

Figure 13.4 Alimentary canal (gut)

butter or margarine to **fatty acids and glycerol**. All of these different substances must now be absorbed and transported round the body.

Absorption in the small intestine

The main function of the small intestine is to absorb the end products of digestion through its wall and then pass them into the circulatory system. The small intestine is very efficient at this job because of its structure (see Figure 13.5). It is very **long** and its internal surface is **folded** and bears thousands of tiny, finger-like projections called **villi** (singular villus). As a result, the small intestine presents a **large absorbing surface area** to the digested food.

Role of villi

The internal structure of a **villus** is shown in Figures 13.6 and 13.7. In addition to presenting a large surface area, villi are ideally suited to the jobs of absorption and transport of digested food because each villus:

- is covered by a cellular lining that is only **one cell thick**. The soluble end products of digestion are therefore able to pass through rapidly.
- contains a dense network of **blood capillaries** into which glucose and amino acids pass. These end products of digestion are then carried by the blood transport system to the liver.
- contains a tiny lymphatic vessel called a **lacteal** that collects fatty acids and glycerol, the products of fat digestion, and passes them into the lymphatic system.

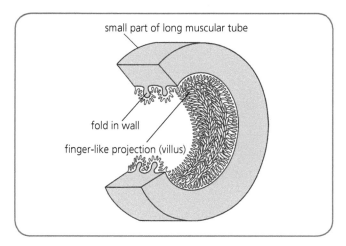

Figure 13.5 Structure of the small intestine

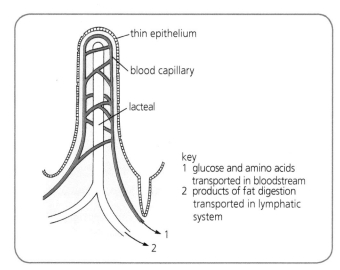

Figure 13.6 Structure of a villus

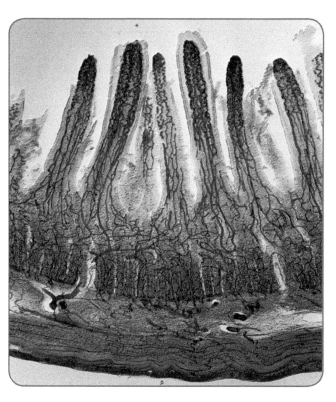

Figure 13.7 Microscopic view of villi

Testing Your Knowledge

1 a) What is an *alveolus*? (1)
  b) Copy and complete Table 13.1 (which refers to features of alveoli) using the following answers: *alveolar wall is thin; to pick up and transport oxygen; inner alveolar surface is moist.* (2)

2 a) State THREE ways in which the structure of the small intestine creates a large surface area suited to the efficient absorption of digested food. (3)
  b) What feature of the lining epithelium of the small intestine enables digested food to pass through easily? (1)

3 Explain how the internal structure of a villus is suited to the transport of digested food. (2)

Feature	Function
alveoli present a large surface area	to absorb much oxygen
	to allow oxygen to dissolve
	to allow oxygen to diffuse through into blood easily
network of tiny blood vessels surrounds alveoli	

Table 13.1

What You Should Know Chapters 11-13

alveoli	guard	pulmonary
amino acids	haemoglobin	red
aorta	hairs	sieve
area	heart	stomata
capillaries	lacteal	sugar
carbon	lignin	thick
cava	low	transpiration
companion	lymphocytes	veins
coronary	mesophyll	ventricles
dead	oxygen	villi
dioxide	phagocytes	water
epidermis	phloem	wider
exchange	pressure	xylem

1 Plants need a transport system to carry _____ to the leaves for photosynthesis and a second transport system to carry _____ from the leaves to all other parts of the plant.

2 Water and mineral salts enter a plant by its root _____. These materials are transported up through the plant in _____ cells (vessels), which are supported by _____ and are _____.

3 Some water arriving in the leaves is used by _____ cells for photosynthesis.

4 Much of the water is lost as water vapour through tiny holes in the leaf _____ called _____. The opening and closing of each stoma is controlled by two _____ cells. The loss of water by evaporation from leaves is called _____.

5 The site of sugar transport in a plant is the _____ tissue. It is alive and composed of _____ tubes and _____ cells.

6 In mammals, the blood circulatory system transports nutrients, _____ and carbon _____ round the body.

7 The _____ is a muscular pump made up of four chambers, the right and left atria and right and left _____.

8 The vena _____ carries blood from the body's tissues and organs to the heart. Blood is transported to the lungs in the _____ arteries and back from the lungs in the pulmonary _____.

9 The _____ carries blood from the heart to the body's tissues and organs. The first branch of the aorta is the _____ artery, which supplies the heart wall itself with oxygenated blood.

10 Arteries carry blood at high _____. They have _____ muscular walls and a narrow central channel. Veins carry blood at _____ pressure and have thinner walls and a _____ central channel.

11 _____ are tiny, thin-walled vessels that present a large surface _____ in tissues and organs, allowing materials to be exchanged.

12 _____ blood cells are specialised to carry oxygen. They contain the red pigment _____.

13 Two types of white blood cell are _____, which engulf pathogens by phagocytosis, and _____, some of which produce antibodies to destroy pathogens.

14 The lungs contain many tiny air sacs called _____. These are thin-walled and present a large surface area in contact with blood capillaries allowing _____ of oxygen and _____ dioxide to occur.

15 The wall of the small intestine bears tiny projections called _____, which are thin-walled and present a large surface area for the absorption of digested food. Each villus contains a blood capillary that absorbs glucose and _____ and a _____ that absorbs the end products of fat digestion.

Applying Your Knowledge and Skills Chapters 7–13

1 Figure KS2.1 shows a complete cycle of the stages that occur before, during and after mitosis in a certain type of animal cell.

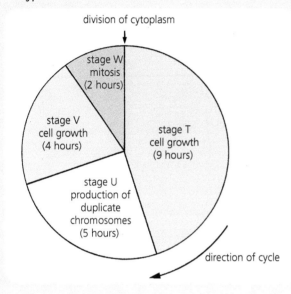

Figure KS2.1

a) Which of the bar graphs in Figure KS2.2 best represents the information in the pie chart? (1)

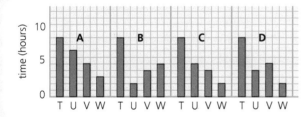

Figure KS2.2

b) How much time was spent on cell growth in one complete cycle? (1)
c) Express your answer to b) as a percentage of the total time required for one cycle. (1)

2 The graph in Figure KS2.3 shows the number of bacteria growing in nutrient broth kept at constant optimum temperature over a period of 30 hours.

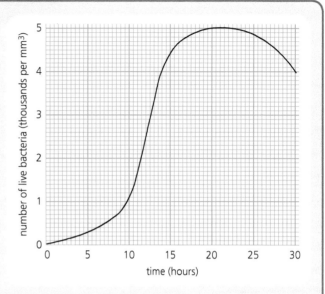

Figure KS2.3

a) i) How many bacteria were present per mm³ at five hours?
   ii) From this point in time onwards, how many more hours passed before the bacteria had doubled in number? (2)
b) During which of the following periods of time was the growth rate greatest? (1)
   A 0–5 hours
   B 5–10 hours
   C 10–15 hours
   D 15–20 hours
c) For how many hours was the number of live bacteria found to be above 4000 per mm³? (1)
d) Give TWO possible reasons why the number of bacteria began to decrease after 23 hours. (2)

3 'A red blood cell is an example of a cell whose structure can easily be related to its function.' Justify this statement. (2)

4 Copy and complete Table KS2.1 overleaf, which refers to structures present in the human body. (9)

System	Organ(s)	Function
digestive	stomach, intestines	
excretory		elimination of wastes
	lungs, trachea	
	ovaries, testes	production of sex cells
nervous	brain	

Table KS2.1

5 The eight statements shown in Figure KS2.4 were made by people being interviewed about the use of human embryos for stem cell research.

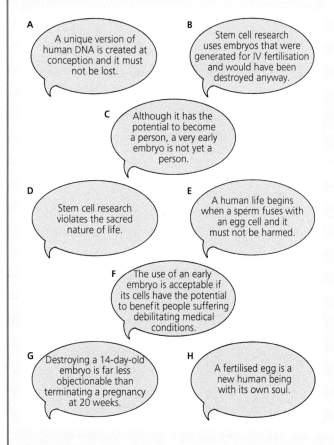

Figure KS2.4

Classify the statements into two groups:
a) those in favour of this type of research (1)
b) those opposed to this type of research. (1)

6 Figure KS2.5 shows the reflex action that happens when the foot comes into contact with a sharp tack.

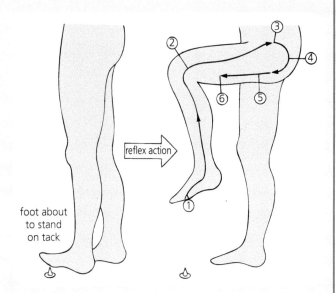

Figure KS2.5

a) Match numbers 1 to 6 with the following statements:
   A Leg muscles contract, removing foot from danger.
   B Impulse enters spinal cord.
   C Impulse travels through motor neuron.
   D Pain receptors in sole of foot are stimulated.
   E Impulse passes through spinal cord.
   F Nerve impulse travels up sensory neuron. (1)
b) Explain why this response is protective. (1)
c) When a piece of dirt lands in the eye, it makes the eye water. Describe this reflex action by identifying the *stimulus, receptor, effector, response* and *protective value*. (5)

7 An experiment was set up to investigate the effect of a pituitary hormone called somatotrophin. The pituitary gland was surgically removed from each of 30 rats of equal mass and age. The rats were divided into two groups, P and Q. Those in P were given daily injections of somatotrophin; those in Q were used as the control. The results are shown in Table KS2.2.

Time (days)	Mean mass (g)	
	group P	group Q
0 (first injection)	220	220
40	240	230
80	250	220
120	280	220
160	320	230
200	390	220
240	410	220
280	410	210
320	430	220
360	440	210
400	440	220

Table KS2.2

a) Present the data as a line graph. (4)
b) Make a generalisation about the mean body mass of the control group of rats during this investigation. (1)
c) What conclusion can be drawn about the normal role of the hormone somatotrophin? (1)
d) i) Why were rats of equal mass and age used?
   ii) Identify TWO other factors that would need to be kept constant in this investigation. (3)
e) Suggest TWO ways in which the reliability of the results could be increased. (2)

8 Figure KS2.6 shows half of a flower.

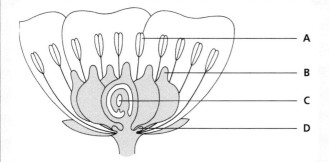

Figure KS2.6

a) Which letter indicates the location of an egg cell? (1)
b) Which letter shows the site of pollen grain production? (1)
c) Which letter indicates the point where a zygote could be formed? (1)

9 Figure KS2.7 represents the life cycle of human beings.

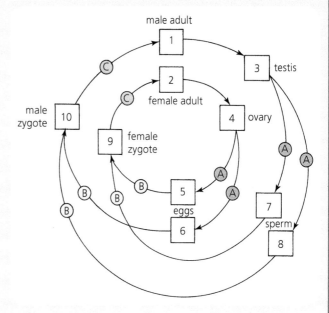

Figure KS2.7

a) Match boxes 1, 2, 3, 4, 5 and 7 with the descriptions haploid or diploid. (2)
b) Identify the process that occurs at all the arrows marked A. (1)
c) What process takes place when two arrows marked B meet? (1)
d) Name the process that occurs at each arrow marked C. (1)

10 a) Which ONE of the following is an example of discrete variation among humans? (1)
   A waist circumference
   B male or female gender
   C length of hand span
   D breadth of foot
b) Which ONE of the following is an example of continuous variation amongst humans? (1)
   A intelligence
   B ear lobe type
   C wavy or straight hair
   D blood group

Type of forefinger print	Number of forefinger prints	Percentage number of forefinger prints
arch	90	
compound	45	
loop	630	
whorl	135	

Table KS2.3

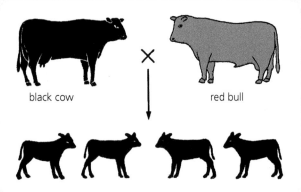

black cow ✕ red bull

Figure KS2.8

c) Table KS2.3 gives the results from a survey of forefinger print types found amongst the students in a college.
   i) Calculate the percentage number of each type of forefinger print.
   ii) Present these percentages as a bar chart. (4)
d) Under the headings *discrete variation* and *continuous variation*, list TWO ways in which identical twins reared in different environments:
   i) would definitely be exactly alike
   ii) could differ from one another. (4)

11 The cross shown in Figure KS2.8 involves the gene for coat colour in cattle.
a) State the two phenotypes of coat colour in this cross. (2)
b) Which form of the gene for coat colour is dominant? (1)
c) Using symbols of your own choice draw a keyed diagram to represent the cross. (2)
d) i) When the calves reach sexual maturity, will they be true-breeding?
   ii) Explain your answer. (2)

12 The data in Table KS2.4 refer to a series of experiments in which a large, leafy shoot attached to a bubble potometer (see Figure 11.7 on page 93) was subjected to various conditions of several environmental factors.
a) Consider each of the following pairs of experiments in turn and for each pair draw a conclusion about the effect of a named factor on the time taken by the bubble to travel 100 mm:
   i) 1 and 2
   ii) 3 and 7
   iii) 6 and 10. (3)
b) Why is it impossible to draw a valid conclusion about the effect of an environmental factor on the time taken by the bubble to travel 100 mm from a comparison of experiments 3 and 8? (1)
c) Which experiment in Table KS2.4 should be compared with experiment 5 to find out about the effect of darkness on time taken by the bubble to travel 100 mm? (1)

		Experiment number									
		1	2	3	4	5	6	7	8	9	10
Environmental factor	wind speed (m/s)	0	0	0	0	15	15	15	15	15	15
	temperature (°C)	5	25	5	25	5	25	5	25	5	25
	air humidity (%)	75	75	95	95	75	75	95	95	75	75
	light (L)/ dark (D)	L	L	L	L	L	L	L	L	D	D
Time taken by bubble to travel 100 mm		3 min 3 s	1 min 35 s	4 min 28 s	3 min 2 s	1 min 56 s	0 min 30 s	3 min 22 s	1 min 57 s	24 min 10 s	22 min 4 s

Table KS2.4

d) Which TWO experiments should be compared in order to find out about the effect of wind speed on time taken by the bubble to travel 100 mm when the plant is in light at 25 °C and in air at 95% humidity? (1)

e) Which TWO experiments should be compared in order to find out about the effect of temperature on transpiration rate by the plant when exposed to wind speed of 15 m/s, in light and in air of 75% humidity? (1)

13 Ringing is the process by which a band of outer tissues containing the phloem is peeled away from the stem of a plant. Before being ringed, the plant in Figure KS2.9 was left in bright light for two days and then leaves A and B were tested for the presence of starch. Next the plant was ringed and kept in darkness for 24 hours. Then leaves A and B were tested again for starch. The results are shown in Table KS2.5.

a) Account for the presence of starch in leaves A and B after two days of bright sunlight. (1)

b) Starch, on being converted to soluble sugar, can be transported from a leaf to other parts of the plant. In which tissue does this transport of sugar occur? (1)

c) Explain why starch is absent from leaf B after 24 hours of darkness. (1)

d) Account for the presence of starch in leaf A after 24 hours of darkness. (2)

	Leaf A	Leaf B
result of starch test after two days in light	+	+
result of starch test after two days in dark	+	−

Table KS2.5

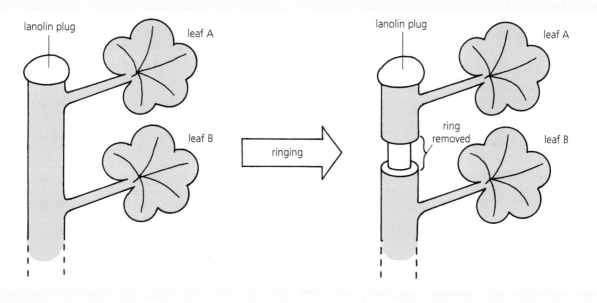

Figure KS2.9

14 The relative concentrations (in units) of carbon dioxide in the air in an alveolus and in the blood in a pulmonary capillary are shown in Figure KS2.10.

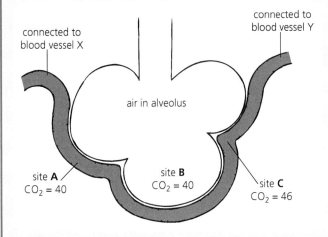

Figure KS2.10

a) At which site is the blood:
   i)   deoxygenated?
   ii) oxygenated?
   iii) Explain your choice. (3)

b) Match the following relative concentrations (in units) of oxygen with sites A, B and C:
   i)   40
   ii) 95
   iii) 100. (2)

c) Match X and Y with the two main vessels that connect the lungs with the heart. (2)

d) The following list gives the substances and layers through which a molecule of oxygen passes on its way from being part of alveolar air to entering a red blood cell. Put them into the correct order. (1)
   1) blood plasma
   2) capillary wall
   3) film of moisture
   4) fluid-filled space
   5) thin alveolar lining

(Since this group of questions does not include examples of every type of question found in SQA exams, it is recommended that students also make use of past exam papers to aid learning and revision.)

3

Life on Earth

14 Ecosystems

A **species** is a group of organisms whose members are able to interbreed with one another and produce fertile offspring. **Biodiversity** means the **total variation** that exists among all living things on Earth. So far, scientists have studied about 1.75 million different species and many more await discovery and investigation. Scotland alone is thought to possess about 50 000 different land and freshwater species and about 39 000 species in the surrounding seas (see Figure 14.1).

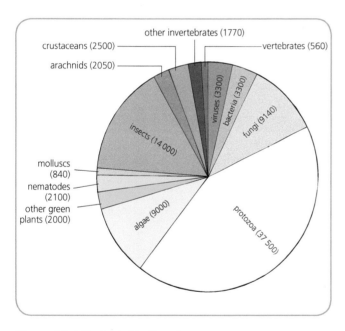

other invertebrates (1770)
crustaceans (2500)
arachnids (2050)
vertebrates (560)
viruses (3300)
bacteria (3300)
fungi (9140)
insects (14 000)
molluscs (840)
nematodes (2100)
other green plants (2000)
algae (9000)
protozoa (37 500)

Figure 14.1 Scottish biodiversity

Producers and consumers

The energy needed by the living things in an **ecosystem** (see page 118) comes from the **sun**. Green plants are called **producers** because they are able to produce their own food by photosynthesis using light energy. Animals (and non-green plants such as fungi) cannot produce their own food from sunlight. They are called **consumers** because they must consume plants or other animals in order to obtain energy.

Among animals, there are different types of consumer. **Herbivores** eat plant material only. **Carnivores** eat animal material only. **Omnivores** eat a mixture of plant and animal material.

Energy flow in a food chain

A **food chain** is a relationship where one type of organism feeds on the previous one in the series and in turn provides food for the next one. A food chain begins with a **producer** (a green plant able to produce food by photosynthesis). Each arrow in a food chain indicates the direction of **energy flow**. When the plant is eaten by an animal (the **primary consumer**), energy is transferred from the plant to the animal. When this first animal is eaten by a second animal (the **secondary consumer**) energy is again transferred and so on through a series of organisms, as shown in Figure 14.2.

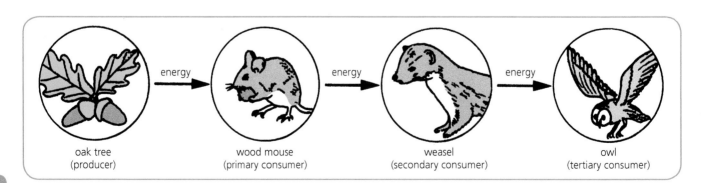

oak tree (producer) → energy → wood mouse (primary consumer) → energy → weasel (secondary consumer) → energy → owl (tertiary consumer)

Figure 14.2 Food chain

Food web

A food chain rarely exists in isolation in nature because the producer is normally eaten by several different consumers which are in turn preyed upon by several different predators. Under natural conditions, an ecosystem really contains many interconnecting food chains. This more complex relationship is called a **food web** (see Figure 14.3).

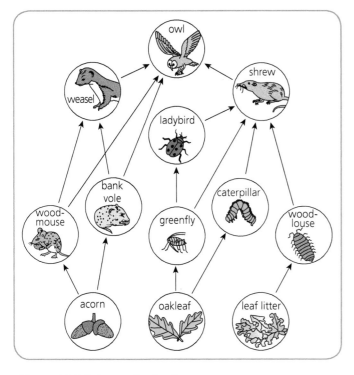

Figure 14.3 Oak tree food web

Disturbing a food chain or web

In a balanced food web a certain mass of green plant material grows continuously and supports a fairly constant and large number of primary consumers, which are in turn consumed by a fairly constant but smaller number of secondary consumers, and so on. However, this balance is disturbed if one of the species is removed from the ecosystem.

Number of links

If a food web has only a **few links** then the effect of removing one species can be severe. Figure 14.4 shows a food web where rabbits are the main source of food for foxes and birds of prey. In 1954–55, the disease myxomatosis wiped out almost the entire population of rabbits. As a result, more tree seedlings and grass grew, but many more lambs than normal were attacked by foxes and birds of prey.

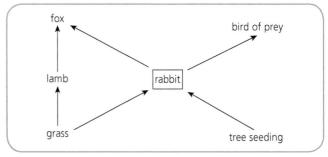

Figure 14.4 Disturbing a food web with few links

If a food web has **many links** the removal of one species may not have such a drastic effect. For example, removal of the limpets from the food web shown in Figure 14.5 would leave more large seaweeds and small algae for other consumers but would not seriously alter the ecosystem.

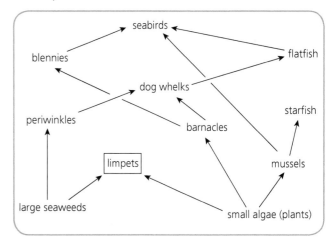

Figure 14.5 Disturbing a food web with many links

Testing Your Knowledge 1

1 Define the terms *species* and *biodiversity*. (3)
2 a) Why are green plants called *producers*? (1)
  b) Why are animals called *consumers*? (1)
3 Distinguish between the terms *herbivore, carnivore* and *omnivore*. (3)
4 a) What is meant by the term *food chain*? (2)
  b) What is a *food web*? (1)
5 a) Predict the effect on the plants if the thrushes are removed from the food web shown in Figure 14.6 overleaf. (1)
  b) Explain your answer. (2)

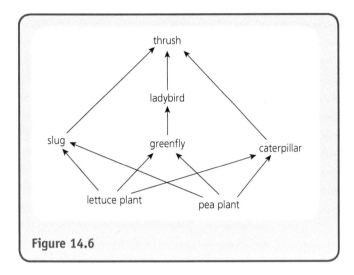

Figure 14.6

Ecosystem

An **ecosystem** is a natural biological unit of the environment (such as a forest). It consists of two closely interrelated parts:

- the entire **community** of living organisms (plants, animals and micro-organisms) that live in **habitats** (such as burrows and tree branches)
- the non-living components and **abiotic factors** (see page 125) present in that area with which the living things interact and by which they are affected.

Put more simply, an ecosystem consists of a **living community** and its **physical environment**.

Ecosystems range in size from very small (such as a freshwater pond – see Figure 14.7) to very large (such as Caledonian forest – see Case Study opposite).

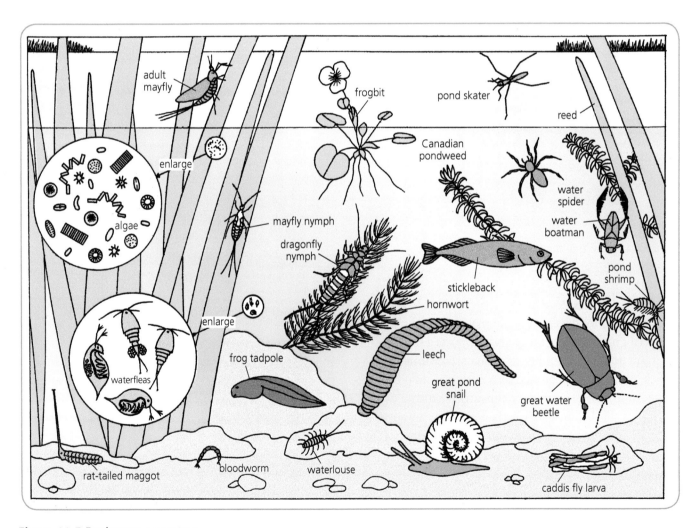

Figure 14.7 Freshwater ecosystem

Case Study Caledonian forest ecosystem

Caledonian forest is a type of ecosystem that once covered large parts of Scotland at the time when trees recolonised the UK after the last ice age. This type of forest is dominated by Scots pine trees but also contains other species of tree such as birch, oak and rowan and smaller plants such as ferns, mosses and lichens.

The 1% of the original native pinewood that survives to the modern day is one of the UK's last remaining wildernesses and it is home to some of the country's rarest wildlife. Some species of bird such as the capercaillie (see Figure 14.8) and Scottish crossbill (see Figure 14.11 overleaf) depend on the Caledonian forest ecosystem as a breeding site and normally fail to breed successfully elsewhere in the UK.

Some birds, such as the black grouse and the long-eared owl, are rarely found living in ecosystems other than the Caledonian forest. This **unique ecosystem** is also the home of many types of mammal including pine marten, red deer, wildcat, red squirrel and European beaver (recently reintroduced to Scotland). Much of the Caledonian forest ecosystem that remains is protected by law. Furthermore, several projects are under way to regenerate and restore more of the forest in the future.

Figure 14.8 Capercaillie in Caledonian forest

Case Study   Heather moorland

About 12% of Scotland is covered by moorland, much of it dominated by heather plants. A **heather moorland** ecosystem (see Figure 14.9) is characterised by low-growing vegetation that thrives on uncultivated acidic soil in conditions of high rainfall, low temperatures and exposure. In addition to heather, the moor contains a variety of other plants such as grasses, cotton-grass and mosses. It also supports a varied animal community such as red deer, red grouse, lapwing, golden plover, hen harrier (predator of red grouse), adder and many insects.

Management
Traditionally heather moorland has been managed for grouse-shooting and much of the land would

Figure 14.9 Heather moorland ecosystem

eventually revert to woodland if it were not **burned** and **grazed**. The moor is burned in patches every 7–15 years so that it is made up of:

- some areas where tall, unburnt heather plants provide **cover** for red grouse
- some areas where recently burnt heather plants produce **new shoots** as food for adult grouse.

Both fire and limited grazing by sheep maintain the moorland's open character. In addition, sheep droppings act as an energy source for insects that are, in turn, eaten by red grouse chicks.

Decline
In recent years, grouse numbers have decreased and many moorland estates have run at a loss. Instead of being carefully managed for shooting and limited grazing, some moorlands have become coniferous forest plantations or have been overgrazed by sheep and red deer. When moorland is overgrazed, many heather plants are replaced by coarse unpalatable grasses and bracken (see page 121). This leads to **reduced biodiversity** among the animal community. Therefore the future of some regions of heather moorland is threatened and they will only survive if the land is carefully managed for grouse-shooting and low-level grazing by sheep.

Niche

An organism's **niche** (more correctly, ecological niche) is the role that it plays within a community. This refers to its **whole way of life** and includes:

- the use that it makes of the resources available in its ecosystem, for example light and nutrients
- its interactions with other members of the community, which may involve factors such as competition (see page 121) and predation (see page 125)
- its tolerance of environmental factors, for example extremes of temperature.

A red fox, for example, is a nocturnal predator. Its food includes small mammals, amphibians, insects and fruit, which it finds in the forests, meadows and river banks that make up its natural habitat. The uneaten remains of the fox's prey provide food for members of the community that act as scavengers and decomposers. The fox's live body, in turn, provides a habitat and food supply (blood) for various parasitic insects.

In recent years, the 'urban' red fox has become very successful by adapting to life in many cities and towns. Its niche is different from that of its 'country cousin' in that its food consists largely of materials discarded by humans and its habitat has become city gardens – a constant source of annoyance to many city dwellers who have no prior experience of interacting with the red fox and who regard their gardens as their own private territory.

Related Activity

Investigating examples of niches

Niche of wild cat

The **wild cat** (see Figure 14.10) is a rare Scottish mammal. Normally its habitat is an inaccessible mountainside or a secluded wooded place among rocks. It is a **nocturnal predator** most active at dawn and dusk because its sight and hearing are its primary senses when hunting. (It has a poor sense of smell.) Its main food is mountain hare, grouse and rabbit and it will sit and wait patiently above a rabbit's warren for its prey to emerge. When food is scarce it will eat any small mammal or bird that it can catch in addition to fish and insects.

The wild cat has few natural enemies, though pine martens may take its kittens. Its main adversaries are human beings, who have wiped it out almost to extinction because of its tendency to create havoc amongst lambs and poultry. The wild cat's niche also includes playing host to various parasitic worms, ticks and fleas.

Niche of Scottish crossbill

Crossbills are parrot-like finches with a beak in which the upper and lower halves cross over one another. The **Scottish crossbill** (see Figure 14.11) is a rare species of crossbill. Experts estimate that there are only about 2000

Figure 14.10 Wild cat

Figure 14.11 Scottish crossbill

in the population. They are found in the Caledonian forest and nowhere else. They live and breed in this coniferous forest where their niche includes competing with other species of crossbill for resources. The females make use of thin twigs, needles and moss to build untidy nests among the evergreen foliage of Scots pine trees where they lay batches of two to five eggs.

The Scottish crossbill is a **specialist feeder** and depends on pine, larch and spruce cones for their seeds. It only eats other seeds when cones are scarce. The bird's unusual beak shape is an adaptation that helps it to extract seeds from cones while the cones are still attached to the tree. The crossed bill is used to wrench off the cone's scales. The bird then extracts the seed using its tongue.

The Scottish crossbill may help to disperse some of the conifer seeds around the forest, especially when parent birds are collecting many seeds to feed their young. The bird's natural enemies include the pine marten, which raids nests and eats eggs and nestlings.

Niche of bracken

Bracken (see Figure 14.12) is the most common fern in the UK. It plays host to a varied community of animals. For example, it provides **food** for the larvae of many butterflies and moths and a **habitat** for the sheep tick, which carries the micro-organism responsible for Lyme disease. Several types of bird including the skylark are found to nest in bracken and use it for **cover**. Adders may even be found basking on bracken on a sunny day. In the absence of woodland, bracken often provides a **canopy** for shade plants such as bluebell and wood anemone and several types of woodland fungi.

Figure 14.12 Bracken

However, bracken produces and releases **poisonous chemicals**. This characteristic enables it to dominate an ecosystem (especially during regrowth after a fire), prevent tree seedlings from growing and choke out rival plants. Normally bracken is not trampled down by cattle or horses because it is poisonous to these animals and farmers try to prevent them from coming into contact with it.

Bracken is a **versatile plant** that can tolerate a range of soil pH from 2.8 to 8.6 (though it does not thrive on marshy soil). It survives adverse conditions by means of **underground stems**, some of which act as storage organs and others as collections of frond (leaf)-forming buds. It is such a successful plant that in many areas it has managed to spread from its traditional hillside habitat to heather moorland, grassland and wooded areas. Livestock farmers regard its continued invasion of pastureland as a major problem.

Competition in ecosystems

Competition occurs in an ecosystem whenever two or more members of the community need a particular **resource** that is in **short supply**. For example, green plants may compete for light, water and soil nutrients; animals may compete for water, food and nesting sites.

Interspecific and intraspecific competition

Interspecific competition occurs when members of **different** species compete for the same resource(s) in an ecosystem. **Intraspecific** competition occurs when the members of the **same** species compete for exactly the same resources in an ecosystem. Intraspecific competition is therefore more intense than interspecific competition because the competing individuals have identical needs.

Evolutionary effect of intraspecific competition

Intraspecific competition regulates the size of the population of the species affected. The weaker members are weeded out by **natural selection** (see page 166).

Related Activity

Investigating interspecific competition

When two different species occupy the same ecological niche, competition may become so fierce that one species tends to force the other out.

Squirrels

The introduction of the North American **grey squirrel** (see Figure 14.13) to Britain has resulted in the widespread decline (almost to extinction) of the **red squirrel**. Both types of squirrel occupy a similar ecological niche in the woodland ecosystem. The grey squirrel is thought to have become so widely distributed and successful because it competes aggressively for food and is able to make use

of a wide variety of foodstuffs including acorns (despite their high tannin content). The grey squirrel is therefore continuing to populate areas at the expense of the more timid red squirrel whose digestive system cannot cope with large quantities of acorns and other seeds rich in tannins.

Trout

A similar situation is developing in some river ecosystems. **Rainbow trout** (see Figure 14.14) that have been introduced from North America are invading the river habitat of the native **brown trout**. The American fish are more aggressive and greedy for food. The fierce competition that has resulted may force the brown trout into decline.

Figure 14.13 Red squirrel and grey squirrel

Figure 14.14 Rainbow trout

Related Activity

Investigating intraspecific competition in plants

Plants of the same species have exactly the same growth requirements. When grown together, they will be in direct competition with one another if any resource is limiting and this competition will be intense.

Oak trees

The oak trees shown in Figure 14.15 are in direct competition for water, soil nutrients and light. The young plant in the centre of the picture is being shaded out by its elders and is unlikely to survive in this example of intraspecific competition.

Cress seedlings

The experiment shown in Figure 14.16 is set up to investigate the effect of density on germinating cress

Figure 14.15 Intraspecific competition in oak trees

seeds. Carton A, with only 100 seeds, represents low density of planting; carton B, with 500 seeds, represents high density. After five days, almost all the seeds in A are found to have successfully germinated and grown into healthy seedlings. Although most of the seeds in B also germinate, many fail to grow into healthy plants and remain yellow and sickly. Of the seedlings that do grow successfully, most are found to be smaller than their counterparts in A.

From this experiment it is concluded that the spaced out plants in carton A grow well because each plant receives an adequate supply of each growth requirement. In carton B where the seedlings are densely packed, much of the light that might have reached each plant is intercepted by the leaves of other plants so that on average each plant receives less light for photosynthesis. In addition their rooting systems become interwoven and may be competing for water.

If a plant is short of water, its stomata stay closed for a longer time and carbon dioxide uptake needed for photosynthesis is reduced. Thus intraspecific competition between densely populated plants results in many individuals growing more slowly.

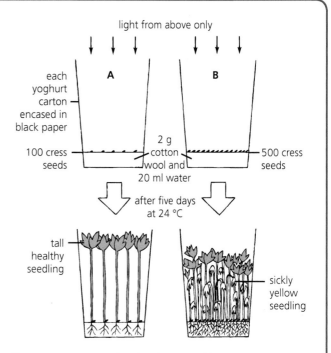

Figure 14.16 Investigating intraspecific competition

Related Activity

Investigating intraspecific competition in animals

Territorial behaviour

Intraspecific competition in animals often takes the form of **territoriality**. This is the name given to behaviour that involves competition between members of the same species (especially birds) for territories. An animal's total **range** is the area that it covers during its lifetime. Within this range a male animal often establishes and inhabits a smaller area called its **territory**, which contains enough food for himself and eventually a mate and their young.

Figure 14.17 Territorial behaviour in robins

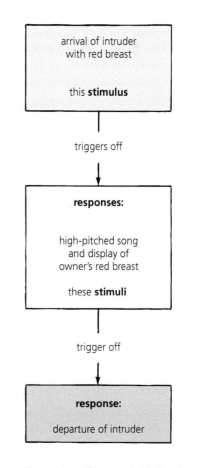

Figure 14.18 Flow chart of territorial behaviour

Robin

Robins defend their territory fiercely using **social signals** (sign stimuli), as shown in Figures 14.17 and 14.18 on the previous page. They are most aggressive at the centre of their territory. The further they move away from the centre, the less likely they are to attack intruders.

Red grouse

The red grouse (see Figure 14.19) lives on moorland and feeds on the shoot tips and flowers of heather plants. Young heather plants are more nutritious than older ones. Rather than compete directly for food, the male red grouse claims a territory large enough to provide food for his dependants during the breeding season.

Figure 14.19 Red grouse

Territorial size varies depending on the availability of food. Each enclosed space in Figure 14.20 represents a red grouse's territory on the same piece of moorland over a period of four years. During years 1 and 4, food was plentiful and the birds only needed to defend small territories. During years 2 and 3, the heather was poor and a larger territory was required to supply a bird's needs. When times were lean, intraspecific competition was more intense and weaker birds that failed to establish a territory did not breed.

Advantages of territorial behaviour

Once territories have been established, aggression between neighbours is reduced to a minimum and **energy** is saved. Territorial behaviour **spaces out** the population in relation to the available food supply. This ensures that there will be enough food for the number of young produced.

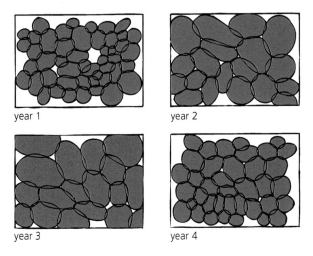

Figure 14.20 Sizes of red grouse territories

Testing Your Knowledge 2

1 a) Of what TWO interrelated parts does an ecosystem consist? (2)
   b) What is the difference between an organism's *community* and its *niche*? (2)
2 a) Under what conditions does competition between organisms occur? (1)
   b) Name TWO resources for which:

   i) plants and
   ii) animals could be competing. (4)

3 a) Explain the difference between *intraspecific* and *interspecific* competition. (1)
   b) Explain why competition between members of the same species is likely to be more intense than that between members of two different species. (1)

15   Distribution of organisms

Biotic and abiotic factors

Factors affecting a species that are directly related to, or are the result of, activities of **living** things are called **biotic** factors. These include, for example, amount of available food, effect of grazing, degree of predation, incidence of disease and level of competition with other species for essential resources.

Non-living factors such as temperature, rainfall, moisture content of soil, light intensity, concentration of oxygen and pH (see Appendix 2) are called **abiotic** factors.

A living organism is only able to survive in a certain habitat and play its part in an ecosystem if a combination of biotic and abiotic factors suited to its needs is present there.

Predation

A **population** is a group of individuals of the same type that forms part of an ecosystem's community. **Predation** is a biotic factor that affects the numbers of both the population of a **prey** animal and the population of its **predator** because a delicate balance exists between the two. (Also see Related Activity below.)

Related Activity

Interpreting predator–prey interaction graphs

The graph in Figure 15.1 shows that an increase in number of prey (perhaps due to climatic conditions favouring growth of their plant food) leads to an increase in predation. As the size of the prey population decreases, competition between predators for the remaining prey becomes more and more intense until eventually the number of predators drops. This in turn allows the prey population to build up again, which leads to a corresponding increase in the predator population and so on. In this series of events, the predator curve takes the same shape as that of the prey but lags behind it since time is required for each change to take effect.

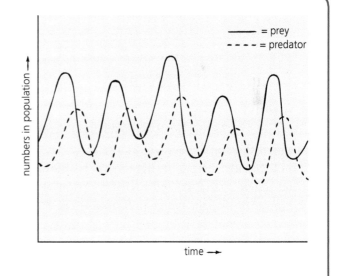

Figure 15.1 Predator-prey interactions

Related Activity

Analysing barn owl pellets

Barn owls are unable to digest the fur and bones of their prey. Once swallowed, these materials are bound into dark-coloured **pellets** in the bird's gut. The owl then gets rid of the pellet by regurgitation ('throwing it up'). An

analysis of the pellets littering the ground near a barn owl's roost allows scientists to identify the bird's prey.

A pellet (see Figure 15.2 overleaf) is dissected using forceps and a magnifying lens while wearing disposable gloves. When the bones present in the pellet are identified (by using, for example, information available on the Barn Owl Trust's website) they are found to have belonged to small mammals such as shrews, voles and mice, which an owl usually swallows whole. Less commonly, a pellet may contain the remains of a rat, a frog or another bird. The bar graph in Figure 15.3 shows the percentage of different prey types present in an analysis of a large number of pellets.

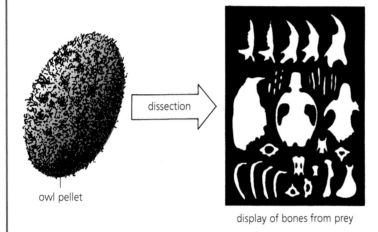

Figure 15.2 Dissection of an owl pellet

Figure 15.3 Analysis of barn owl pellets

Measuring abiotic factors

Light intensity

Light intensity can be measured using the **light meter** part of a light/moisture meter (see Figure 15.4). The switch on the meter is set at the light meter position. The meter is held so that the **light sensitive panel** is directed towards the light source to be measured. The reading is taken from the light intensity scale once the pointer has stopped moving. All measurements are taken as near the same time as possible during a period of constant light intensity so that a comparison between readings at different locations is valid.

Soil moisture content

Soil moisture content can be measured using the **moisture meter** part of a light/moisture meter. The switch on the meter is set at the moisture meter position. The **moisture probe** is pushed into the soil to a depth of 40 mm. The reading is taken from the moisture meter scale once the pointer has stopped moving. The probe is wiped with a tissue between samples to prevent moisture from one reading affecting the next one.

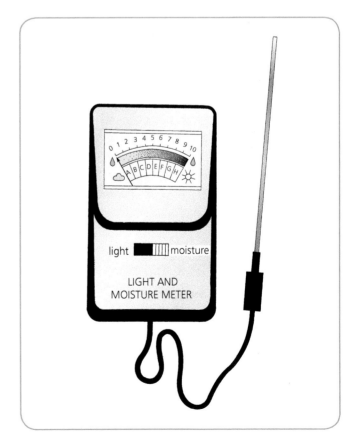

Figure 15.4 Light/moisture meter

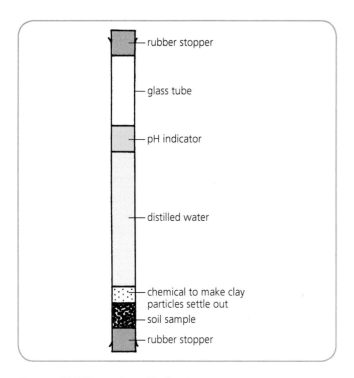

Figure 15.5 Measuring pH of soil

Temperature

A **thermometer** or a **temperature probe** can be used to measure the temperature of soil or of pond water, for example. The probe may be linked to appropriate data-logging software on a computer.

pH

A **pH meter** can be used to measure the pH (acidity or alkalinity) of soil (or pond water). The meter's probe is inserted into the soil and it is connected to data-logging software on a computer. Alternatively the pH of soil can be tested using the apparatus shown in Figure 15.5. The tube is shaken vigorously and then allowed to settle. A clear zone of liquid appears and its colour is compared with a chart of pH colours to identify the soil's pH.

Testing Your Knowledge 1

1 a) In general, what is the difference between a *biotic* and an *abiotic* factor that can affect biodiversity? (2)
   b) Organise the following factors into two groups – biotic and abiotic: pH, competition, food supply, light intensity, predation, temperature, disease, grazing, rainfall. (2)

2 a) Describe how a light/moisture meter could be used to measure:
      i) light intensity   ii) soil moisture content. (4)
   b) Identify a precaution that should be adopted in each case so that a comparison between readings at different locations is valid. (2)

Sampling the organisms in an ecosystem

It is rarely possible to count all of the plants and animals in an ecosystem because this would take too long and would probably cause permanent damage to the ecosystem. Instead, small **samples** that represent the whole ecosystem are taken. To make the procedure **valid**, each sampling unit must be **equal in size** and be chosen at random (so that the results are not biased). From the samples an estimate of an organism's **abundance** for the whole ecosystem can be calculated.

Quantitative techniques for measuring biotic factors

The use of **pitfall traps** and **quadrats** are two of the many different quantitative techniques for sampling and counting the organisms in a particular ecosystem.

Figure 15.6 Pitfall trap

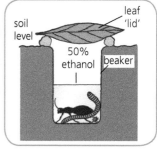

Figure 15.7 Improved pitfall trap

Pitfall trap

Table 15.1 overleaf summarises the use of a **pitfall trap** for trapping and counting the numbers and types of animal present in a soil ecosystem such as the one shown in Figure 15.8.

Ecosystem	Sampling technique	Possible sources of error	Ways in which errors may be minimised
soil	trapping using a **pitfall trap** (see Figure 15.6) (this operates on the basis that animals that are active on the soil surface will fall into the trap and be unable to climb out again)	the numbers and types of animals may not be representative of the ecosystem as a whole (this source of error is true of all sampling techniques)	set up several traps (replication makes the results more reliable)
		birds may eat trapped animals	disguise the opening with a lid such as a leaf supported on sticks (see Figure 15.7)
		some animals may eat the others	check traps regularly or put preservative liquid such as 50% ethanol in the bottom of the traps (see Figure 15.7)

Table 15.1 Pitfall trapping technique

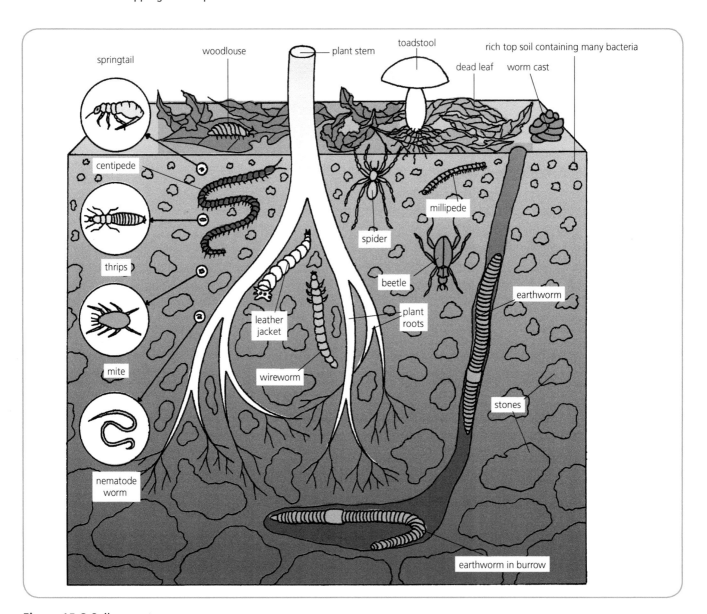

Figure 15.8 Soil ecosystem

Quadrats

A further method of sampling uses quadrats. A **quadrat** is a rectangular-shaped sampling unit of known area. Its frame is normally made of wood, metal or string. This form of sampling is often used to estimate the numbers of plants in an ecosystem (or slow-moving animals such as shellfish on a rocky shore).

Estimating the total number of thistles in a field

Figure 15.9 shows a quadrat that encloses an area of 1 square metre. The sites for several quadrats (say ten) are chosen at random (see Figure 15.10) and the number of thistles present in each quadrat is counted. From this information the average number of thistles per square metre is calculated. The length and breadth of the field is measured and its area calculated. An estimate of the total number of thistles in the field is then worked out.

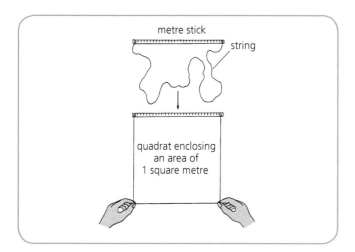

Figure 15.9 Quadrat

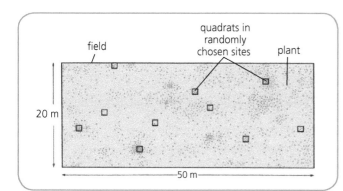

Figure 15.10 Sampling a field

Example of calculation

Table 15.2 shows a typical set of results.

Quadrat	Number of thistles
1	11
2	0
3	5
4	15
5	1
6	7
7	3
8	7
9	9
10	4

Table 15.2 Quadrat results

Total number of thistles in ten quadrats = 62

Average number of thistles per quadrat (i.e. per m^2) = 62/10 = 6.2

Total area of field (length × breadth) = 50 m × 20 m = 1000 m^2

Estimate of total number of thistles in field = 6.2 × 1000 = 6200 thistles

Possible sources of error

Number of quadrats

Ten quadrats may be too small a number of samples to give a fair representation of the number of plants present in the ecosystem. This is especially likely if the plant being investigated tends to grow in clusters. Most or all of the quadrats might just happen to land in positions where none of the plants under investigation grow (see Figure 15.11).

In or out?

Some of the type of plant being considered might be located partly inside and partly outside a quadrat. Should they be included in the count or not?

Ways in which errors may be minimised

A much larger number of quadrats could be studied by the class working as, say, ten groups with each group doing ten quadrats and the whole class pooling their results. This would improve the **reliability** of the results and therefore give a more accurate representation of the number of plants present.

A basic rule could be established and followed by everyone involved (see Figure 15.12). Any plant partly in and partly out falling on the bottom or left-hand side of the quadrat counts as in, whereas any plant on the top or right-hand side counts as out.

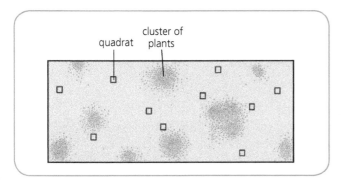

Figure 15.11 Too few quadrats

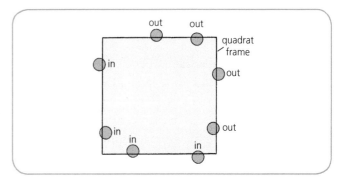

Figure 15.12 In or out?

Related Activity

Measuring abundance of plant types

The **abundance** of a plant type is a measure of the extent to which it occurs in an environment. For example, one type of plant might be found to occur frequently and another only rarely in an ecosystem. Abundance can be measured using the type of quadrat shown in Figure 15.13.

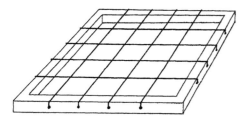

Figure 15.13 Frame quadrat

The quadrat is placed at randomly chosen sites in the ecosystem being studied. The number of squares that contain the type of plant being investigated is counted (up to a maximum of 25). This method does not involve counting the number of plants.

Figure 15.14 shows an example. The abundance of plantains = 3 (out of 25), dandelions = 5, buttercups = 8 and grass = 24.

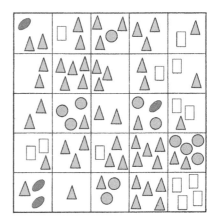

= plantain
= dandelion
= buttercup
= grass

Figure 15.14 Measuring abundance

Related Information

Further techniques for biotic factors

Use of a Tullgren funnel

A **Tullgren funnel** (see Figure 15.15) is used to extract tiny animals that live in the soil's air spaces. The animals in the soil sample move down and away from the hot, dry, bright conditions created by the light bulb and fall through the sieve into the water below.

Pond netting

The **net** (see Figure 15.16) is moved rapidly through the water, catching animals that are quickly transferred to screw-top jars containing pond water. When investigating the bottom of the pond, the handle of the net is rotated as shown in Figure 15.17. This keeps the net closed on the way down and on the way back up again. It prevents entry of surface animals and escape of animals collected at the bottom.

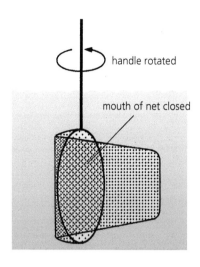

Figure 15.17 Careful use of net

Tree beating

A branch of a tree is held over a tray and given a few sharp taps with a **walking stick**, as shown in Figure 15.18. Small organisms that fall onto the tray are transferred to a large plastic bag. Back in the laboratory, a **pooter** (see Figure 15.19) is used to trap animals for close inspection.

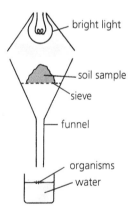

Figure 15.15 Tullgren funnel

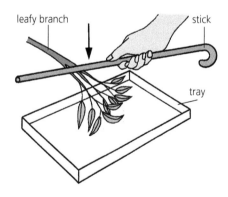

Figure 15.18 Tree beating

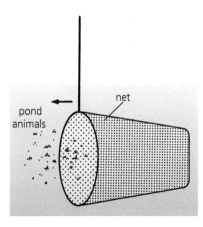

Figure 15.16 Net

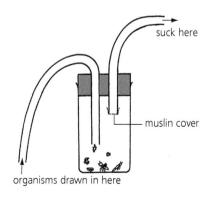

Figure 15.19 Pooter

Related Activity

Identifying the organisms collected

Each organism can be identified by comparing it to pictures in a suitable book. However, this method can take a very long time. A much better method is to use a **key**. A key may be branched or in the form of paired

statements. Figures 15.20 and 15.21 give keys to animals commonly found in soil and pond ecosystems. (After identification, organisms should be returned to their natural habitats so that as little disturbance as possible is caused to the ecosystem.)

1	body has no legs	go to 2
	body has legs	go to 6
2	body not divided into sections (segments)	go to 3
	body divided into segments	go to 5
3	body worm-like	**nematode worm** (a)
	body not worm-like	go to 4
4	shell present	**snail** (b)
	no shell present	**slug** (c)
5	no more than 13 segments present	**fly maggot** (d)
	more than 13 segments present	**earthworm** (e)
6	6 jointed legs present	go to 7
	more than 6 jointed legs present	go to 11
7	grub-like insect	go to 8
	adult insect	go to 9
8	non-jointed legs present on abdomen	**caterpillar** (f)
	non-jointed legs absent from abdomen	**beetle larva** (g)
9	thin waist between thorax and abdomen	**ant** (h)
	no thin waist between thorax and abdomen	go to 10
10	spring attached to abdomen	**springtail** (i)
	no spring attached to abdomen	**beetle** (j)
11	8 legs present	go to 12
	more than 8 legs present	go to 13
12	body divided into 2 parts	**spider** (k)
	body not divided into 2 parts	**mite** (l)
13	14 legs present	**woodlouse** (m)
	more than 14 legs present	go to 14
14	each body segment has 1 pair of legs	**centipede** (n)
	each body segment has 2 pairs of legs	**millipede** (o)

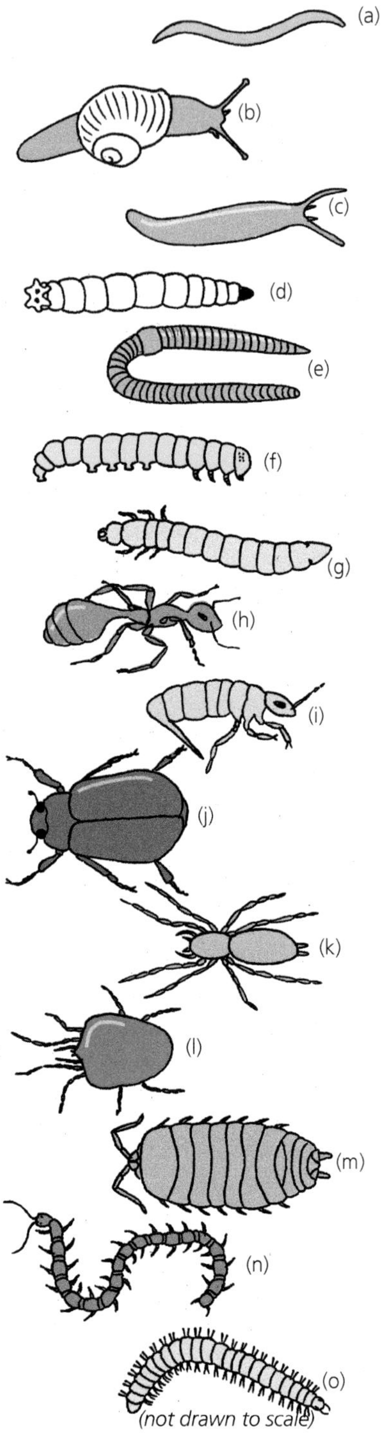

(not drawn to scale)

Figure 15.20 Key to soil animals

1	shell present	go to 2
	shell absent	go to 3
2	shell made of 2 halves	**freshwater mussel** (a)
	shell made of 1 part	**pond snail** (b)
3	body not divided into segments	go to 4
	body divided into segments	go to 5
4	body threadlike	**nematode worm** (c)
	body flat	**flatworm** (d)
5	no obvious jointed legs present	go to 6
	jointed legs present	go to 8
6	feelers and tail hooks present	**midge larva ('bloodworm')** (e)
	no feelers or tail hooks present	go to 7
7	suckers present at both ends	**leech** (f)
	no suckers present	**freshwater worm ('redworm')** (g)
8	6 legs present	go to 9
	more than 6 legs present	go to 13
9	lives in protective tube	**caddis fly larva** (h)
	free living	go to 10
10	2 pairs of wings present	go to 11
	no wings present	go to 12
11	large dark body, threadlike feelers present	**water beetle** (i)
	small light body, no threadlike feelers present	**water boatman** (j)
12	2 tail appendages ('prongs') present	**stonefly** (k)
	3 tail appendages ('prongs') present	**mayfly nymph** (l)
13	8 legs present	go to 14
	more than 8 legs present	go to 15
14	small body in 1 part	**water mite** (m)
	large body in 2 parts	**water spider** (n)
15	body flattened downwards like a woodlouse	**waterlouse** (o)
	body flattened sideways, gills on front legs	**freshwater shrimp** (p)

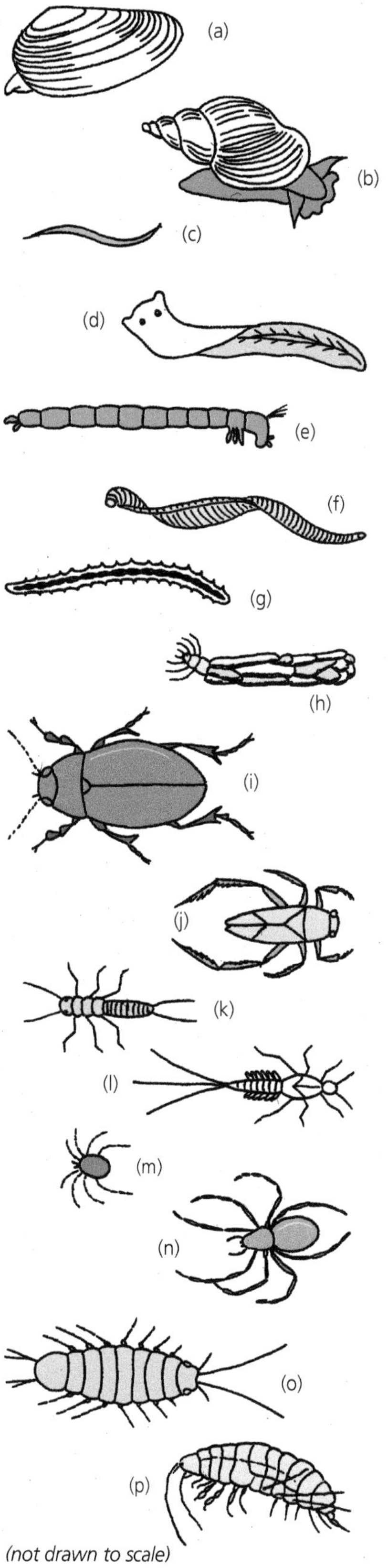

(not drawn to scale)

Figure 15.21 Key to pond animals

Effect of biotic and abiotic factors on distribution

Both biotic and abiotic factors can affect the distribution of a population of organisms in an ecosystem. In the Related Activity below, the effect of light intensity (an abiotic factor) on the distribution of daisies is investigated.

Related Activity

Investigating the effect of light intensity on distribution of daisies

In this investigation the area being studied contains a grassy field and part of an oak wood. On a sunny day the abiotic factor (light intensity) shows a wide range of variation from high in the middle of the field to low in the wood. To investigate if light intensity affects the distribution of daisy plants, a **line (belt) transect** is set up, as shown in Figure 15.22. A length of string is pegged from point X in the field to point Y in the wood. The string has been marked in advance at 1 metre intervals, giving ten sample sites.

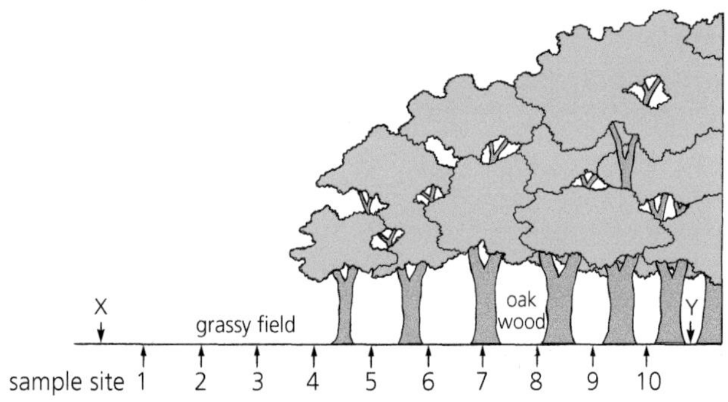

A quadrat of the type used to measure abundance (see Figures 15.13 and 15.14 on page 130) is placed at each sample site and the number of squares containing daisies is counted. The light meter part of the meter shown in Figure 15.4 is used to measure the light intensity at each sample site. Table 15.3 gives a typical set of results. The results show that the daisies are most abundant in light of high intensity and become less frequent as light intensity decreases.

Possible mechanism determining distribution of daisies

Daisies are green plants that make their food by photosynthesis using light energy. In conditions of low light intensity, it is possible that they are unable to obtain sufficient light energy to produce food. They are therefore unable to survive in the dimly lit conditions in the oak wood. This investigation suggests that the abiotic factor, **light intensity**, determines the distribution of daisies. However, it is possible that other abiotic factors such as soil moisture content or soil pH could also be playing a part in their distribution.

Figure 15.22 Line transect

Sample site	1	2	3	4	5	6	7	8	9	10
Abundance of daisies (score out of 25)	19	20	18	21	12	8	4	1	0	0
Light intensity (A = low, H = high)	H	H	H	H	G	F	E	D	C	C

Table 15.3 Abundance of daisies

Effect of biotic and abiotic factors on biodiversity

The biodiversity present in an ecosystem is affected by biotic and abiotic factors. Some factors may increase biodiversity but many bring about its decline.

Effect of grazing on biodiversity

Grazing is a biotic factor that can affect species diversity in a grassland. Natural grassland normally contains a rich variety of plant species. Some types of grass are especially sturdy and show vigorous growth; others are more delicate.

Rabbits are relatively unselective grazers. Their effect on a grassland's plant community depends on the level of grazing pressure applied. At low levels of grazing pressure, the aggressive dominant grasses are not held in check and tend to drive out the less vigorous plants. Species diversity among the plant community under these circumstances is low.

As grazing pressure increases (see Figure 15.23) more and more of the competitive dominant grasses are eaten and kept in check. This tends to free resources and space for the smaller, more delicate plant varieties and promotes an **increase in biodiversity** of species. However, at high intensities of grazing pressure, species **biodiversity decreases** again as the dense population of rabbits struggle to find food and drive some of the more delicate species within the plant community to extinction.

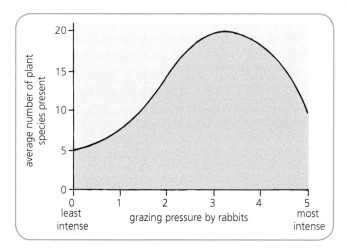

Figure 15.23 Effect of grazing on species diversity

Effect of pH and temperature on biodiversity

pH and **temperature** are abiotic factors that can affect the variety of fish species in an ecosystem.

pH and acid rain

The combustion of coal by electricity-generating power stations produces sulphur dioxide and nitrogen oxides, which combine with water to form acids. When present in excess in the atmosphere, they lead to the formation of **acid rain**, which has a devastating effect on several types of ecosystem and their communities.

Figure 15.24 shows the effect of the decrease in pH of the water in a loch on the variety of fish species found in this aquatic ecosystem. The length of each bar represents the lower range of pH that could be tolerated by the species of fish.

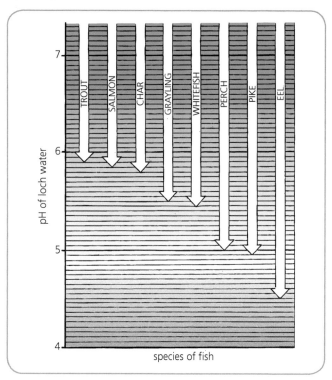

Figure 15.24 Effect of pH on the variety of fish species

Temperature and thermal pollution

Some types of electricity-generating power station use local river water as a coolant. When the water is returned to the river, it is considerably warmer and causes **thermal pollution**. The increase in temperature of river water is accompanied by a decrease in its dissolved oxygen content (see Figure 15.25) and a

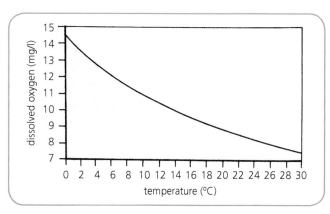

Figure 15.25 Effect of temperature on dissolved oxygen content of water

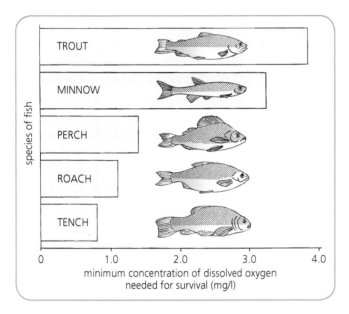

Figure 15.26 Effect of oxygen concentration on the variety of fish species

decrease in the variety of fish species in the river (see Figure 15.26). Whereas all of the species shown in the diagram could be present in water containing 4 mg/l of oxygen, only tench would be found at 1 mg/l of oxygen.

Indicator species

Some species of living thing only thrive well under certain environmental conditions such as very clean air, badly polluted water and so on. Such organisms that indicate the quality of the environment are called **indicator species**. They are important because their presence (especially in large numbers) shows that an ecosystem is affected by a particular set of environmental conditions.

| **Research Topic** | **Lichen survey** |

Lichens (see Figure 15.27) are simple plants composed of a fungus and an alga, often found growing on the bark of trees and on rock surfaces. Different species of lichen act as an indicator species because they vary in their sensitivity to **sulphur dioxide (SO_2)**, a gas that causes atmospheric pollution. The variety of lichen species present in an ecosystem decreases as the concentration of SO_2 in the air increases (see Figure 15.28).

The data obtained from a **lichen survey** can be used to construct an air pollution map of an area as shown in Figure 15.29 opposite.

Figure 15.27 Lichens

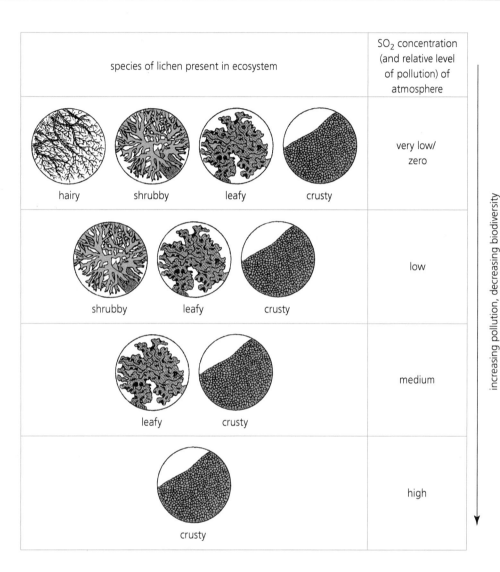

Figure 15.28 Effect of air pollution on variety of lichen species

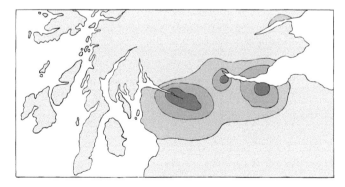

Figure 15.29 Air pollution map based on lichen survey

Effect of oxygen concentration on freshwater invertebrates

Research into the effects of low oxygen concentration on aquatic animals shows that species differ in their ability to tolerate the **varying oxygen levels** found in a river **polluted** with organic waste (see Figure 15.30). Some animals such as **mayfly** and **stonefly nymphs** are commonly found in clean water rich in oxygen. However, they cannot survive when the river's oxygen content is greatly reduced by the presence of a huge number of decomposer micro-organisms.

If the water becomes extremely polluted and contains almost no oxygen, then the only animals that survive are **rat-tailed maggots** and **sludgeworms**. As the water gradually becomes a little less polluted, animals such as **midge larva** ('**bloodworms**') and **water louse** are commonly found because they can survive in water of low oxygen concentration.

When pollution eventually decreases to a fairly low level, water plants can survive and produce oxygen by photosynthesis. The oxygen content of the water begins to rise and **freshwater shrimps** and **caddis fly larvae** are now found to become numerous since they can tolerate low levels of pollution. **Mayfly** and **stonefly nymphs** are only found again when the water is very clean and contains plenty of oxygen.

Indicator species

Since each of the above species of invertebrate animal only thrives within a specific range of oxygen concentration, it acts as an **indicator species**. The relationship of these indicator species to oxygen concentration and level of pollution of the water in their ecosystem is summarised in Table 15.4.

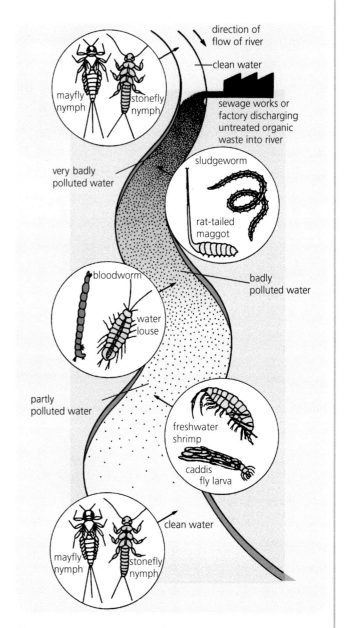

Figure 15.30 Freshwater indicator organisms

Indicator species present	Oxygen concentration of water	Level of water pollution
mayfly nymph stonefly nymph	high	absent or very low
freshwater shrimp caddis fly larva		low/medium
bloodworm water louse		high
rat-tailed maggot sludgeworm	low	very high
no animals present	zero	extreme

Table 15.4 Freshwater indicator species

Testing Your Knowledge 2

1 a) Name a technique that could be used to sample the invertebrate animals living in a garden's soil. (1)
  b) Name the apparatus that you would use and describe how you would use it. (2)
  c) i) Identify a possible source of error when using this technique.
     ii) State how this error could be minimised. (2)

2 a) Name a technique that could be used to estimate the total number of dandelion plants in a large patch of grass. (1)
  b) Describe how you would use the apparatus. (2)
  c) i) Identify a possible source of error when using this technique.
     ii) State how this error could be minimised. (2)

3 a) Define the term *indicator species*. (1)
  b) Give ONE named example of an indicator species and state what it indicates about its environment. (2)

What You Should Know Chapters 14–15

abiotic	interspecific	pitfall
biotic	intraspecific	predation
carnivores	light	quadrats
community	many	quality
competition	meters	sampling
ecosystem	niche	temperature
herbivores	omnivores	valid
indicator	pH	variation

1 An _____ is made up of a community of living organisms and the non-living factors with which they interact. A _____ is the role that an organism plays within a _____ and often involves _____ and predation.

2 Animals that eat plant material only are called _____. Those that eat animal material only are called _____, and those that eat both types of food are called _____.

3 Biodiversity means the total _____ that exists among living things. Biodiversity in an ecosystem can be affected by _____ and abiotic factors that can increase or decrease it.

4 Biotic factors such as grazing and _____ are caused by living things; abiotic factors such as pH and _____ are non-living.

5 When competition for the same resource(s) occurs between individuals of different species in an ecosystem, it is called _____ competition; when it occurs between individuals of the same species it is called _____ competition.

6 Biologists investigate an ecosystem's community by _____ its plants and animals. To do this, they often use quantitative techniques such as _____ traps and _____.

7 For sampling to give a _____ representation of the community, _____ samples must be taken and mean results calculated.

8 An ecosystem is affected by many _____ factors including _____ intensity, temperature, moisture content and _____. These can be measured using appropriate _____ and probes linked to computer software.

9 A species that by its presence indicates the _____ of an environment is called an _____ species.

16 Photosynthesis

Capture of light energy

Photosynthesis consists of a series of enzyme-controlled reactions that allow green plants to make their own food. This process involves the capture of **light energy** from the sun. Light energy is trapped by the green pigment **chlorophyll**. Chlorophyll is found in discus-shaped structures called **chloroplasts** present in green leaf cells (see Figure 1.1 on page 2).

Production of carbohydrate

A **carbohydrate** is a compound containing the chemical elements carbon (C), hydrogen (H) and oxygen (O) combined together using energy. The production of a carbohydrate food (such as **glucose**, a type of sugar) requires a supply of the raw materials **water** and **carbon dioxide** to be available to supply the necessary chemicals. During photosynthesis, molecules of carbon dioxide combine with hydrogen from water in the presence of chlorophyll and light energy to form sugar. Oxygen is released as a by-product of the process.

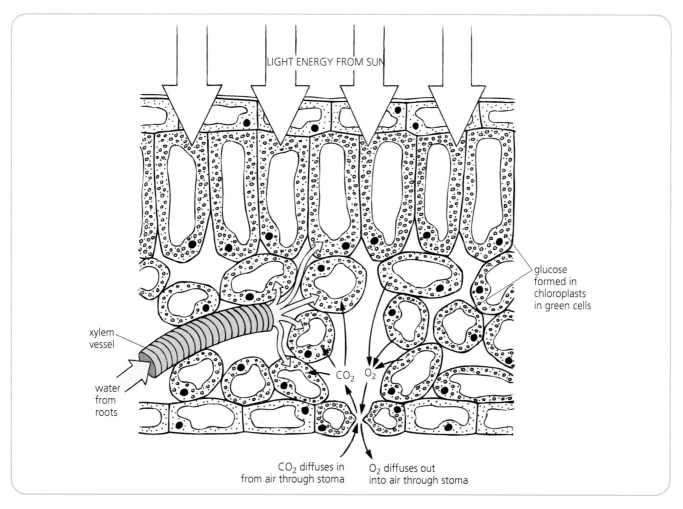

Figure 16.1 Photosynthesis in a leaf of a land plant

Summary

The process of photosynthesis is summarised in Figure 16.1 and by the following equation:

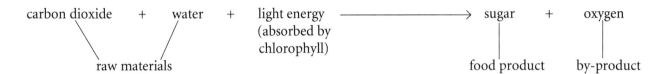

carbon dioxide + water + light energy (absorbed by chlorophyll) $\longrightarrow$ sugar + oxygen

raw materials   food product   by-product

Biochemistry of photosynthesis

Photosynthesis consists of two distinct stages:

- the light-dependent stage consisting of several **light reactions**
- the temperature-dependent stage called **carbon fixation**.

Light-dependent stage

During this series of light reactions, **light** energy is trapped by chlorophyll and converted to **chemical** energy. This is used to generate energy-rich **ATP**. Some energy is also used during this stage to **split water** into hydrogen and oxygen (see Figure 16.2). The oxygen is released as a by-product and diffuses out of the plant. The hydrogen and the ATP are used during the carbon fixation stage of photosynthesis. It cannot proceed without them.

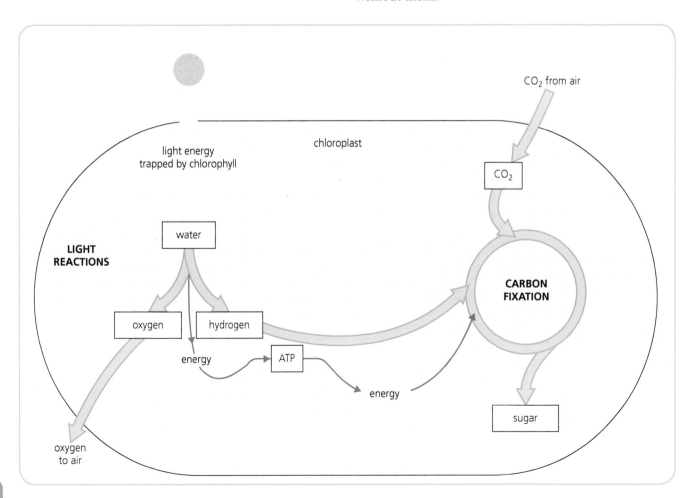

Figure 16.2 Biochemistry of photosynthesis

Carbon fixation

This second stage of photosynthesis also occurs in chloroplasts (see Figure 16.2). It consists of several **enzyme-controlled reactions** in the form of a cycle that result in **carbon** (from carbon dioxide) becoming 'fixed' into a **sugar** (such as glucose) by combining with the hydrogen from the light-dependent stage. ATP supplies the energy needed to drive this process.

Conversion of glucose into plant products

Carbohydrates

As a plant grows, it continues to make glucose by photosynthesis. Some of this sugar is broken down again by **respiration** to supply the plant with energy for vital processes such as cell division and reproduction. Some of the remaining glucose molecules become linked into long chains and packed together into **starch** grains found in the cell's cytoplasm (see Figures 16.3, 16.4 and 16.5). This plant product is the plant's store of food and can be converted back to sugar when energy is required. Starch is therefore called a **storage** carbohydrate.

Other glucose molecules are built into long chains of **cellulose**. These are gathered together to form the fibres needed to build cell walls (see Figure 16.3 and Figure 2.16 on page 14). Cellulose is therefore called a **structural** carbohydrate.

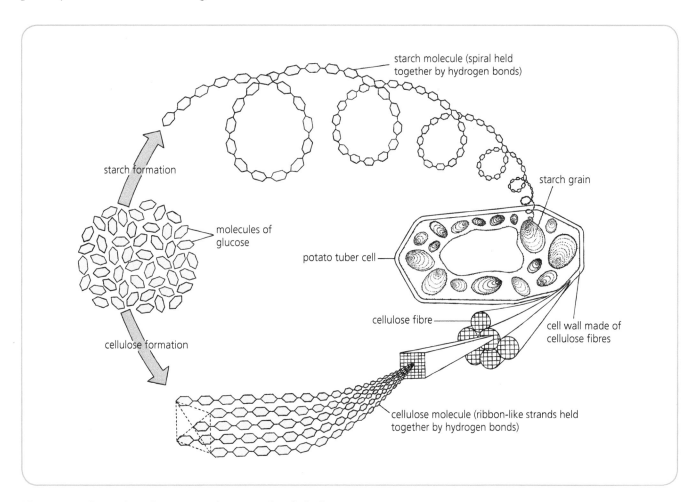

Figure 16.3 Formation of storage and structural carbohydrate

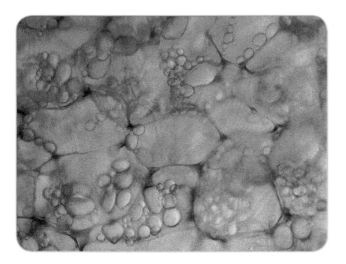

Figure 16.4 Potato tuber cells in water

Figure 16.5 Potato tuber cells in iodine solution

Investigation

Effect of light on starch production

A green plant is set up as shown in Figure 16.6 and left in bright light for two days. With the aid of a cork borer, a disc is then cut from a leaf that has been in light. When the disc is tested by following the procedure shown in Figure 16.7, it is found to turn **blue-black**, showing that it contains starch. When the experiment is repeated using a disc from the leaf that has been in darkness, it does not turn blue-black, showing that it lacks starch. Therefore it is concluded that **light** is necessary for starch production (and photosynthesis).

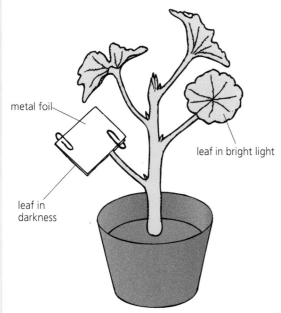

Figure 16.6 Investigating the need for light

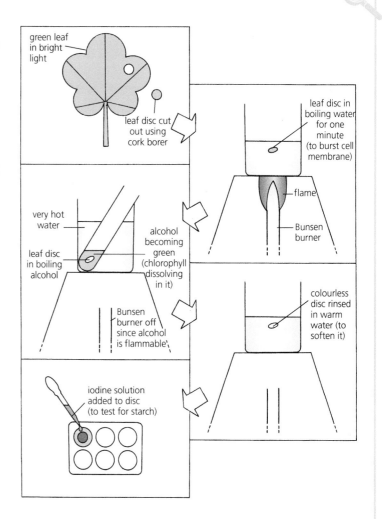

Figure 16.7 Testing a leaf disc for starch

Investigation

Effect of carbon dioxide on starch production

Two destarched green plants are set up as shown in Figure 16.8 and left in bright light for two days. A leaf disc from each plant is then tested for starch by following the procedure shown in Figure 16.7. The disc from plant A is found to contain starch whereas the disc from plant B is found to lack starch. Therefore it is concluded that **carbon dioxide (CO_2)** is necessary for starch production (and photosynthesis).

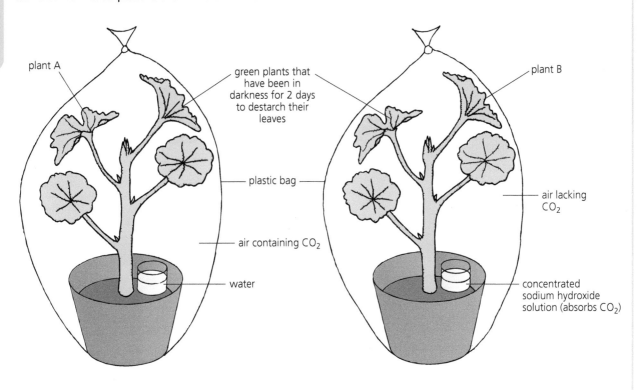

plant A

green plants that have been in darkness for 2 days to destarch their leaves

plastic bag

air containing CO_2

water

plant B

air lacking CO_2

concentrated sodium hydroxide solution (absorbs CO_2)

Figure 16.8 Investigating the need for carbon dioxide (CO_2)

Effect of chlorophyll on starch production

When the starch test is applied to a leaf that is **variegated** (has two colours, one of which is green, as shown in Figure 16.10), only the green regions react positively with iodine solution (see Figure 16.9). Therefore it is concluded that **chlorophyll** is necessary for starch production (and photosynthesis).

Figure 16.10 Plant with variegated leaves

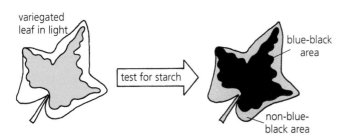

variegated leaf in light

test for starch

blue-black area

non-blue-black area

Figure 16.9 Investigating the need for chlorophyll

145

Testing Your Knowledge 1

1 Figure 16.11 gives an illustrated summary of the biochemistry of photosynthesis.

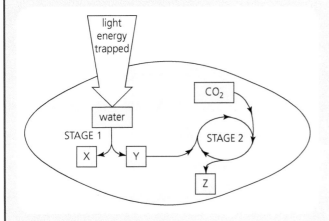

Figure 16.11

a) i) Name the discus-shaped organelle within which photosynthesis takes place.
   ii) Give the colour and name of the chemical substance responsible for trapping light energy. (3)
b) i) Identify substances X, Y and Z.
   ii) Which of these is the carbohydrate end product?
   iii) Which of these is a by-product that diffuses out of the plant? (5)
c) i) Name stages 1 and 2.
   ii) Which high-energy substance needed to drive stage 2 has been omitted from the diagram? (3)

2 Name:
a) the structural and
b) the storage carbohydrate formed in plant cells using glucose molecules made by photosynthesis. (2)

Factors affecting photosynthetic rate

Several environmental factors affect the rate of photosynthesis. These include light intensity, carbon dioxide concentration and temperature.

Effect of varying light intensity

Elodea bubbler experiment

The number of oxygen bubbles released per minute by the cut end of an *Elodea* stem (see Figure 16.12) indicates the rate at which photosynthesis is proceeding. At first the lamp (see Figure 16.13) is placed exactly 100 cm from the plant and the number of bubbles released per minute is counted. The lamp is then moved to a new position (say 60 cm from the plant) and the rate of bubbling noted (once the plant has had a short time to become acclimatised to this new higher light intensity).

The process is repeated for lamp positions even nearer the plant, as recorded in Table 16.1. When this typical set of results is displayed as a graph (Figure 16.14), it can be concluded that as light intensity increases, photosynthetic rate also increases until it reaches a maximum, in this case of 25 bubbles per minute at around 64 units of light.

Figure 16.12 Release of oxygen from cut stem of *Elodea*

Distance from plant (cm)	Units of light (calculated using a mathematical formula)	Number of oxygen bubbles released per minute
100	4	4
60	11	10
40	25	19
30	45	24
25	64	25
20	100	25

Table 16.1 *Elodea* bubbler results

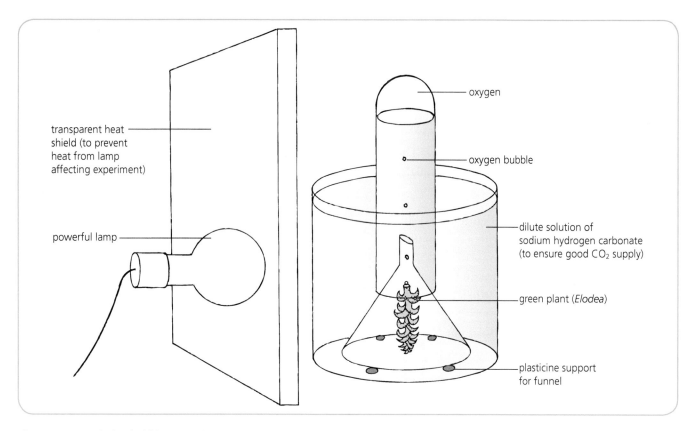

Figure 16.13 *Elodea* bubbler experiment

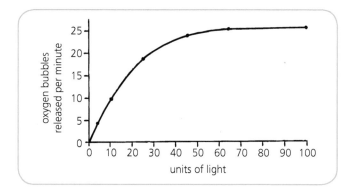

Figure 16.14 Graph of *Elodea* bubbler experiment results

Ingredients needed to make one loaf	Baker A's stock	Baker B's stock
500 g flour	5 kg flour	5 kg flour
30 g fat	60 g fat	300 g fat
10 g yeast	50 g yeast	50 g yeast
5 g sugar	40 g sugar	15 g sugar

Table 16.2 Factors limiting bread making

Limiting factors

Consider the information in Table 16.2. Although Baker A has plenty of flour, yeast and sugar, he can only make two loaves because he has a limited supply of fat. Baker B has plenty of fat but she can only make three loaves because her stock of sugar limits production. A **limiting factor** is a factor that holds up a process because it is in short supply.

Similarly, limiting factors such as light intensity, carbon dioxide concentration and temperature can hold up photosynthesis. In the *Elodea* bubbler experiment described above, further increase in light intensity beyond 64 units does not increase photosynthetic rate. It remains at 25 bubbles per minute. This is because some other factor such as shortage of carbon dioxide is now holding up the process and acting as a limiting factor.

Investigating the effect of varying carbon dioxide concentration

In this experiment the concentration of carbon dioxide made available to the *Elodea* plant is gradually increased by adding appropriate masses of sodium hydrogen carbonate to the water. The number of oxygen bubbles released is counted as before. The lamp is kept in one position to give uniform light of medium intensity.

The graph of a set of results (see Figure 16.15) shows that when the plant is supplied with a carbon dioxide concentration of only 1 unit, the rate of photosynthesis is **limited** by this low concentration of carbon dioxide to 3 oxygen bubbles per minute. When the carbon dioxide concentration is increased to 2 units, rate of photosynthesis increases to 6 bubbles per minute but no further, since carbon dioxide concentration becomes limiting again.

A further increase in carbon dioxide concentration to 3 units brings about a further increase in photosynthetic rate. However, beyond this point the graph levels out and any further increase in carbon dioxide concentration does not affect photosynthetic rate. This is because some other factor (such as light intensity) is now limiting the process.

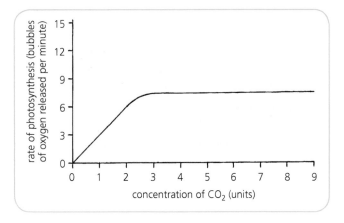

Figure 16.15 Variation in CO_2 concentration at one light intensity

Analysing the effect of carbon dioxide concentration at different light intensities

In Figure 16.16, graph ABC represents a repeat of Figure 16.15 and graph ADE represents the results from a further *Elodea* bubbler experiment using the same plant in conditions of constant high light intensity. In this second experiment, an increase in carbon dioxide concentration to 4 and then to 5 units brings about a corresponding increase in photosynthetic rate in each case. This is because carbon dioxide concentration is still the **limiting factor** up to 5 units of carbon dioxide when the light intensity is high. However, beyond 5 units of carbon dioxide, the graph levels off again since some other factor (such as light intensity or temperature) has become limiting.

Investigating the effect of varying temperature

The apparatus shown in Figure 16.13 is adapted for use in this experiment by using a large water bath whose temperature is under thermostatic control. Plastic bags containing ice cubes are used to create low temperatures. The plant is given light of constant high intensity and a rich supply of carbon dioxide to ensure that neither of these factors limits the process.

The graph in Figure 16.17 shows that the photosynthetic rate, in this case, rises to its maximum level at around 35 °C, the **optimum** temperature for this type of plant. (The optimum temperature varies from species to species and is often lower than 35 °C). At values below the optimum, temperature is acting as the limiting factor. At values above the optimum temperature, the photosynthetic rate drops rapidly. This is because photosynthesis consists of many reactions controlled by enzymes that are denatured at higher temperatures.

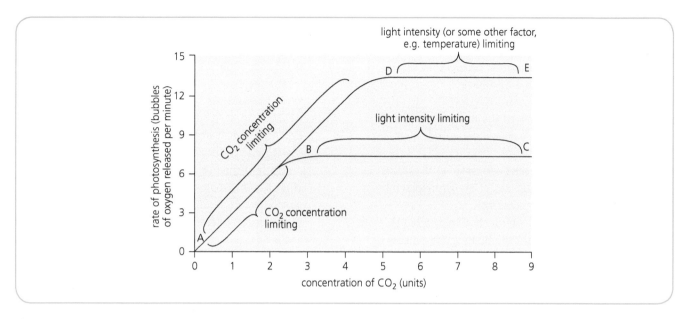

Figure 16.16 Variation in CO_2 concentration at two light intensities

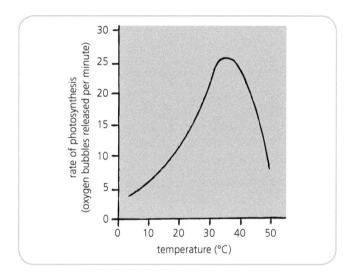

Figure 16.17 Effect of temperature on photosynthetic rate

Impact of limiting factors

If one of the factors that affect the rate of photosynthesis is limiting then the plant makes less sugar than it would under optimum conditions. Cell growth slows down and less food is stored as starch.

If a crop plant in a field is affected in this way there is little that the farmer can do to solve the problem. However, if some factor is limiting the rate of photosynthesis by a crop of plants such as tomatoes in a glasshouse, steps can be taken to improve the situation.

If **temperature** is the only limiting factor then on cold, bright winter days, for example, the temperature of the glasshouse can be increased using a heating system. When **light intensity** is the only limiting factor then on dull winter days, for example, the horticulturist can employ electric lighting to increase light intensity. When **carbon dioxide concentration** is the limiting factor, extra supplies can be added to the air in a glasshouse. This creates a **carbon dioxide enriched environment**, which increases the yield of photosynthetic product. By using a combination of these techniques, the horticulturist can produce early crops and maximise photosynthetic yield.

Testing Your Knowledge 2

1 a) Identify TWO environmental factors that can be demonstrated by the experiment in Figure 16.13 to affect the rate of photosynthesis. (2)

   b) As the distance between the powerful lamp and the *Elodea* plant is gradually decreased, what effect does this have on:
      i) the intensity of the light reaching the plant?
      ii) the number of bubbles of oxygen released by the plant per minute? (2)

2 a) What is meant by the term *limiting factor*? (1)

   b) When no other factors are limiting the process of photosynthesis, what effect on photosynthetic rate is brought about by raising the temperature of the plant from:
      i) 15 °C to 35 °C?
      ii) 35 °C to 55 °C? (2)

3 a) Explain why it is economically viable to supply tomato plants but not wheat plants with a carbon dioxide enriched environment. (2)

   b) What benefit would a horticulturist gain by providing a crop of plants in a glasshouse with supplementary heating and lighting during the winter? (1)

17 Energy in ecosystems

Energy loss

As energy flows through a food chain or web, a progressive **loss** of about 90% occurs at each level for two reasons.

Firstly, an organism uses energy to build its body. However, this may include parts such as cellulose cell walls or bone or skin or horns that, when eaten by the next consumer, may turn out to have little or no nutritional value. These parts tend therefore to be left uneaten by the next consumer or to be expelled undigested as faeces. As a result, energy is lost from the food chain. Some of this energy is, however, gained by the ecosystem's decomposers.

Secondly, most of the energy gained by a consumer in its food is used for **moving** about and, in warm-blooded animals, for **keeping warm**. Much energy is therefore **lost** as heat and only about 10% on average of the energy taken in by an organism is used for **growth** and incorporated into its body tissues. Figure 17.1 shows the fate of energy as it is transferred along a marine food chain. Figure 17.2 shows an alternative way of presenting this information involving the flow and loss of energy in an ecosystem.

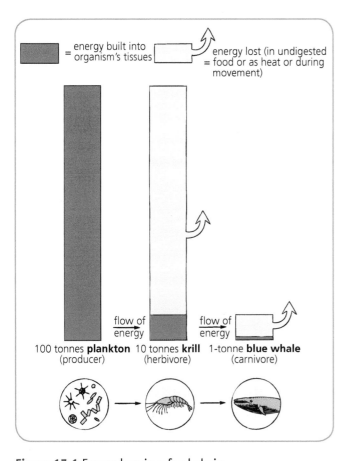

Figure 17.1 Energy loss in a food chain

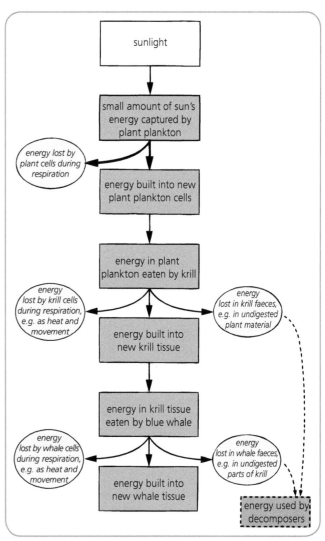

Figure 17.2 Energy transfer and loss

Length of a food chain

More efficient use is made of food plants by humans consuming them **directly** rather than first converting them into animal products, since this cuts out at least one of the energy-losing stages in the food chain.

Pyramid of numbers

Consider the following food chain:

alga → water flea → stickleback → pike

In terms of numbers, the producers (the algae) are found to be the most numerous, followed by the primary consumers and so on along the chain, with the final consumer being the least numerous. This numerical relationship is called a **pyramid of numbers** and is often illustrated in the form shown in Figure 17.3.

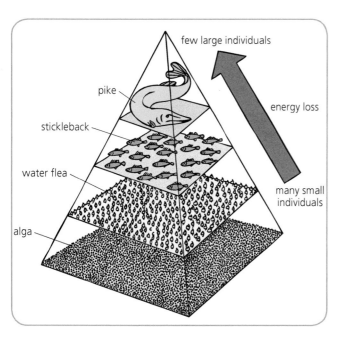

Figure 17.3 Pyramid of numbers

The relationship takes the form of a pyramid because:

- the **energy loss** at each link in the food chain limits the quantity of living matter that can be supported at the next level
- the final consumer tends to be **larger** in body size than the one below it, and so on.

A simpler way of representing a pyramid of numbers is shown in Figure 17.4.

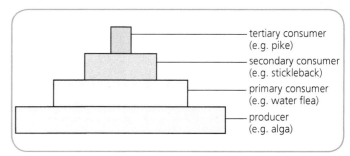

Figure 17.4 Pyramid of numbers (alternative style)

Pyramid of energy

A comparison between the organisms found at the different levels in a food chain can be made based on **productivity**. This is measured as grams of dry mass per square metre per year and then converted into its energy equivalent in kilojoules per square metre per year. The results are used to construct a **pyramid of energy**, which illustrates the energy content at each level and the transfer of energy from one level to another. Figure 17.5 shows a pyramid of energy for a river ecosystem.

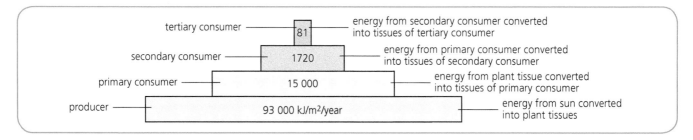

Figure 17.5 Pyramid of energy

Investigating irregular pyramids of numbers

Tree as producer

In some food chains the producer is a **single large plant** and the 'pyramid' therefore takes a different form. Figure 17.6, for example, shows the irregular pyramid of numbers for the food chain:

oak tree → caterpillar → shrew → owl

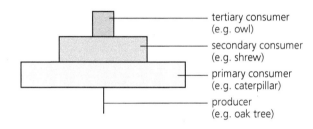

Figure 17.6 Pyramid with large single producer

Parasite as final consumer

In some food chains, the final consumer consists of a **large population of tiny parasites** and the 'pyramid' takes a different form. Figure 17.7 shows the irregular pyramid of numbers for the food chain:

grass → gazelle → lion → flea

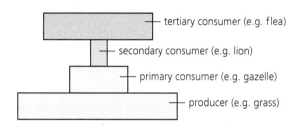

Figure 17.7 Pyramid with a large population of parasites

True pyramid

A pyramid of numbers may take an irregular shape because it is based on different body sizes (see the Related Activity above). However, when any pyramid of numbers, regular or irregular, is represented as a pyramid of **energy**, it takes the form of a **true pyramid** because only a proportion of energy (about 10%) is successfully transferred from one level to the next.

1 Describe TWO ways in which energy may be lost from a food chain. (2)
2 a) Copy the pyramid of numbers shown in Figure 17.8 and complete it using the following organisms: *water flea, pike, alga, stickleback*. (2)
  b) Which of these organisms is the secondary consumer? (1)
  c) Which population of organisms in this pyramid contains most energy? (1)
  d) Give ONE reason why the numbers decrease towards the top of a normal, regular pyramid of numbers. (1)
3 Give the meaning of the term *pyramid of energy*. (1)

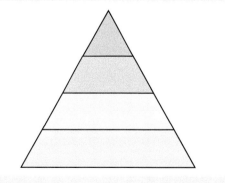

Figure 17.8

What You Should Know Chapters 16–17

ATP	fixation	limited
base	generate	numbers
cellulose	glucose	numerous
chain	heat	photosynthesis
chlorophyll	hydrogen	pyramid
consumer	intensity	split
dioxide	level	temperature
energy	light	transfer

1 Green plants use _____ as their source of energy to make food. This process is called _____. It is a two-stage process.

2 During the first stage, which is made up of light reactions, light energy is captured by green _____ in chloroplasts. Some energy is used to _____ ATP. Some energy is used to _____ water into oxygen and _____.

3 During the second stage of photosynthesis, called carbon _____, hydrogen produced during the first stage is combined with carbon _____ to form glucose. This process requires energy, which is supplied by _____.

4 Some _____ molecules produced by photosynthesis are used in cellular respiration, others are converted to products such as starch and _____.

5 Photosynthesis is affected by environmental factors such as light _____, carbon dioxide concentration and _____. Its rate is therefore _____ by whichever one of these factors is in short supply.

6 About 90% of _____ is lost at each level in a food _____ as undigested material or as _____ or during movement.

7 A pyramid of _____ illustrates the numerical relationship between the organisms in a food chain with the producer (the most numerous at the _____ and the final _____ (the least _____) at the top.

8 A _____ of energy illustrates the energy content at each _____ in a food chain and the _____ of energy from one level to another.

18 Food production

Population growth

Unlike other species, human beings have managed to overcome most of the environmental factors that normally prevent a population in its natural ecosystem from increasing indefinitely in number. Humans have used their exceptional brainpower to:

- remove the threat of predation
- develop vast areas of land for food production
- improve public health.

As a result, the human population continues to increase rapidly and is expected to reach 9 billion by the year 2040 (see Figure 18.1).

Food production

The continuous increase in population size needs to be matched by an equivalent increase in food yield. Attempts to achieve this goal are made by:

- growing vast monocultures of food crops
- employing methods of intensive farming
- developing genetically modified (GM) crops as alternatives to the use of pesticides.

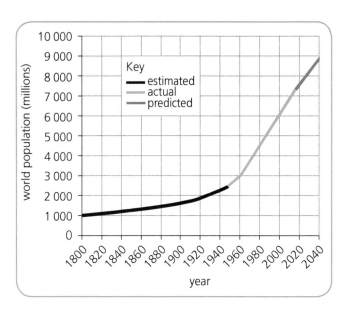

Figure 18.1 Human population growth

Research Topic   Monocultures and intensive farming

Monoculture

A **monoculture** (see Figure 18.2) is a vast cultivated population of one type of crop plant whose members are often genetically identical. In order to feed the ever-growing human population, natural ecosystems have been cleared to accommodate vast monocultures of crops such as wheat, maize, rice and potatoes.

Intensive farming

Compared with traditional smallholder cultivation of arable farmland, **intensive agriculture** enables farmers to produce more food from the same acreage of land. Intensive farming involves:

Figure 18.2 Monoculture – in this case maize

- growing high-yield crop plants as vast **monocultures** and supplying the plants with large quantities of chemical fertiliser containing nitrates, which increase crop yield
- making regular use of chemicals such as **herbicides** (to kill weeds that would compete with the crop and **pesticides** (to remove insects and other pests that would consume the crop and so reduce its yield)
- rearing farm animals indoors and often confining them to restricted spaces (see Figure 18.3). This so-called **'battery' farming** reduces the quantity of energy lost by the animal as heat to the surrounding environment and the energy used by the animal for movement. Therefore it leaves more available for the growth of the animal.

Figure 18.3 Restricted space during 'battery' farming

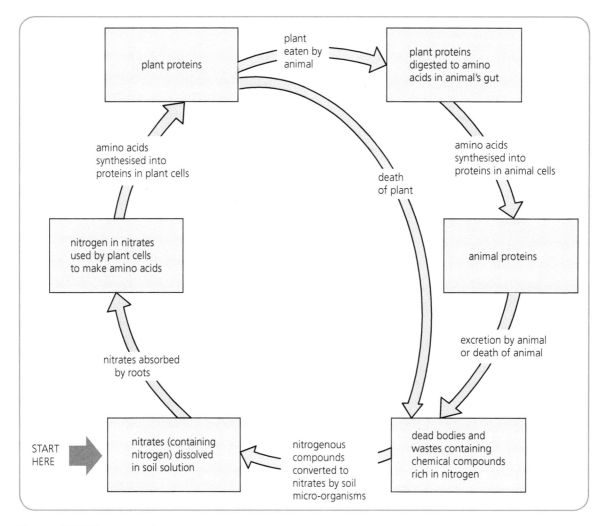

Figure 18.4 Nitrogen cycle

Nitrogen cycle

All living things need nitrogen to make protein. However, plants and animals cannot make direct use of the nitrogen gas present in air. Instead, plants absorb nitrogen from the soil in the form of nitrate. Animals must eat plant or animal protein to obtain their supply of amino acids for protein synthesis. Figure 18.4 shows some of the main processes that occur during the nitrogen cycle. The maintenance of a natural balanced ecosystem depends on the repeated cycling of nitrogen. Natural processes brought about by soil micro-organisms replace nitrates as fast as they are used by plants.

Need for fertiliser

When a crop is grown on cleared land and then harvested and removed, the ecosystem's natural balance is disturbed. The nitrogen cycle is broken since almost no dead plant material is returned to the soil for breakdown by micro-organisms. Less nitrates are generated and the soil becomes less fertile. This problem is tackled by adding fertiliser to the soil to increase its nitrate content in advance of planting the next crop.

Leaching

If fertiliser from fields **leaches** into a river or loch, it makes the water over-rich in mineral nutrients such as nitrates. This promotes the rapid growth of algae (simple water plants), which form an **algal bloom** (see Figure 18.5).

This scummy layer on the surface of the water reduces the quantity of light able to reach the aquatic plants below, which then die of starvation. These dead water plants and dead algae from the bloom act as a food source for aerobic bacteria. As the bacteria bring about the decomposition process, they undergo a population explosion (see Figure 18.6). Their high rate of respiration uses up the **oxygen** dissolved in the water. This reduction in the water's oxygen level has an adverse effect on the river or loch's animals and many die from lack of oxygen.

Figure 18.5 Algal bloom

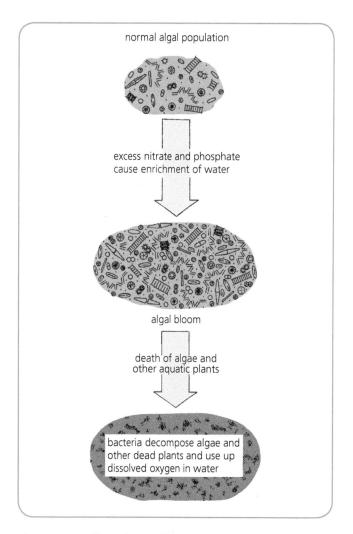

Figure 18.6 Effect of algal bloom on oxygen supply

157

Pesticides

Importance

It is estimated that without pesticides, cereal crop yield would be reduced by 25% after one year and by 45% after three years. It is unlikely that wholesome, healthy, unblemished food, sufficient in quantity to feed the world's ever-increasing population, could be produced without the aid of pesticides.

Adverse effects

In addition to having the desired effect, some pesticides also have unintended effects on other populations in the ecosystem. The extent of these disruptive side-effects depends on various factors highlighted by posing the following questions:

- Is the pesticide **specific** in its action or is it poisonous to a wide variety of organisms?
- Is it **biodegradable** or does it persist for a very long time in the environment?
- Is the concentration that is used kept to a **minimum** or is it used in excess?

In the past people often failed to address these issues. This led to the following sequence of events:

- Molecules of toxic, non-biodegradable chemical in pesticides enter and persist in the bodies of producers.
- The toxic molecules are passed along food chains and become more and more concentrated at each level. (This build-up of toxic substances is called **bioaccumulation**.)
- High concentrations of the toxic chemical accumulate in the tissues of the final consumers, which are seriously affected, sometimes fatally.

Research Topic	**Bioaccumulation of DDT**

The pesticide **DDT**, which is both **persistent** and highly **toxic**, was widely used in the 1950s and 1960s. It was found to pass easily through food chains with the effect described above. In the example in Figure 18.7, the producer becomes contaminated with a very low concentration of the pesticide blown off neighbouring farmland during the spraying of crops. The concentration increases, however, when plant material is eaten by primary consumers and the DDT persists in their cells.

Progression on up the pyramid of numbers and biomass leads to ever-increasing concentrations of the chemical **accumulating** in living cells. Finally, the few large tertiary consumers at the top suffer severe poisoning.

Thin egg shells

The female birds at the top of this affected pyramid suffer an upset in the balance of the hormones that control the manufacture of strong egg shells. As a result, they lay eggs with **thin shells** that often break during incubation, leading to a significant decrease in reproductive success. The graph in Figure 18.8 shows that as few as 25 parts per million of the pesticide can reduce a shell's thickness by as much as 15%. Once the shell is more than 23% thinner than normal, it breaks with disastrous results.

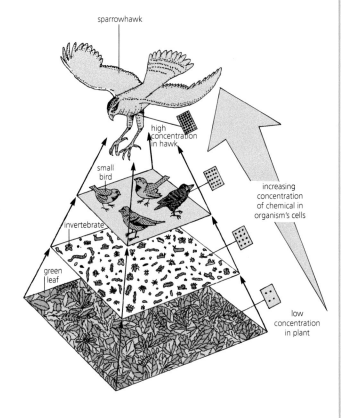

Figure 18.7 Accumulation of chemical in a food pyramid

Aquatic ecosystems

Pesticide sprays can also be washed off farmland into local watercourses. If they are non-biodegradable, they similarly increase in concentration among the members of the freshwater ecosystem's community. So persistent is DDT that it has found its way into marine ecosystems despite the dilution effect of the sea on polluted river water. Traces of DDT have even been found in the fatty tissue of penguins in the Antarctic.

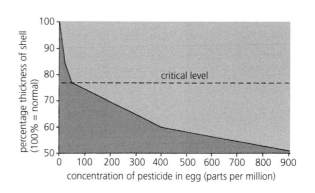

Figure 18.8 Effect of pesticide on thickness of shell

Combating the harmful effects of intensive farming

The significant increase in yield achieved by intensive farming may be accompanied by **harmful effects** such as:

- destruction of natural ecosystems and loss of biodiversity
- pollution of the environment by pesticides
- death of helpful, as well as harmful, insects due to pesticides
- bioaccumulation of pesticides along food chains.

Genetically modified (GM) crops

One way of trying to redress the balance is to use **GM crops** such as maize that produce their own 'built-in' toxin. These GM crops are able to **resist pests** and do not need to be sprayed with pesticide. Other GM crops such as certain strains of rice take up nitrogen more efficiently than their non GM relatives. If used widely these plants would **reduce the need for fertiliser** to be added to the soil.

| **Research Topic** | Genetically modified (GM) crops |

Current levels of food production are not expected to meet the projected future demand and already many people in the world lack sufficient food. One way of increasing food production is to develop **genetically modified (GM)** crops. In recent years genetic engineers have managed to do this by inserting a useful gene from some other organism into the cells of a crop plant (see Chapter 5). This process has produced:

- maize (see Figure 18.9) containing a bacterial gene for a toxin that makes it resistant to insect pests
- potato that is resistant to fungal blight
- cucumber, courgette and pepper that are resistant to disease-causing viruses
- oilseed rape, sugar beet and soya bean that can tolerate herbicide (weedkiller) while nearby weeds are destroyed
- golden rice that contains a source of vitamin A (see page 33)
- apple and tomato varieties that are slow to ripen (see page 34).

Figure 18.9 GM maize

The toxins that some GM crops contain are harmless to humans.

Most GM crops increase food supply by reducing the quantity of crop that is lost. In addition many of them allow farmers to decrease their reliance on pesticides and herbicides without suffering a decrease in yield. Unfortunately there is no guarantee that a GM plant such as maize resistant to corn borers (see page 170) will remain resistant permanently. Eventually a mutant form of the pest may evolve and overcome the plant's resistance. Then a new strain of the GM crop will be required.

Despite the obvious advantages of GM crops, some people believe that these plants may carry some inherent risk to human health. In addition, they are concerned that genetic material introduced into crops could somehow become incorporated into other species in the ecosystem. This, they argue, could produce new, undesirable strains such as 'super-weeds' resistant to weedkillers.

Biological control

A further alternative practice used to mitigate the effects of intensive farming is **biological control**. This is the reduction of a pest population by the deliberate introduction of one of its **natural enemies** such as a predator or a parasite. The use of ladybirds to clear greenfly from plants is an example of biological control. The natural enemy acts as the control agent and specifically targets the pest. No environmental pollution occurs and no chemicals persist and accumulate in food chains.

Related Activity

Investigating biological control

Use of ladybirds

Ladybirds (or more correctly ladybird beetles) prey on **scale insects** and **aphids** (see Figure 18.10). Ladybirds are active from spring to autumn and can consume a large number of aphids such as greenfly in one day. Ladybirds are used by some horticulturists to act as a means of **biological control** to keep their crop of tomatoes or cucumbers in a glasshouse free of aphids. Some gardeners purchase cultures of ladybirds from suppliers to release in their garden (for example, on rose bushes infested with greenfly) in preference to using an insecticide spray.

Use of virus

Myxomatosis is a disease that was first discovered in South America. It is caused by a **virus** and affects the South American cottontail **rabbit** (*Sylvilagus*), which suffers a relatively harmless form of the disease.

Many years ago, the European rabbit (*Oryctolagus*) was introduced into Australia. It lacked competitors and became so successful that by 1950 its numbers had reached 600 million. It was regarded as a serious pest and the myxomatosis virus from South America was deliberately introduced into the rabbit population as a means of **biological control**.

The disease developed by the European rabbit involves swelling of mucous membranes, formation of skin tumours, blindness, fever, pneumonia and death. It had a devastating effect and reduced the rabbit numbers in Australia to 100 million in two years. The rabbits that survived were naturally resistant to the disease and gradually the population recovered. A further attempt at biological control was carried out in 1996 using a second virus (rabbit calcivirus).

Figure 18.10 Ladybird and aphids

Use of cactus moth

The **cactus moth** (*Cactoblastis*) is a native of South America. Its larvae feed on a type of **cactus** whose scientific name is *Opuntia*. In South America, both organisms are members of a balanced ecosystem where their population numbers are held in check naturally.

Prickly pear cactus is a type of *Opuntia*. It was introduced to Australia in 1840. In the absence of its natural enemies and competitors, the cactus flourished. Within a few years it was spreading at a rate of almost half a million hectares a year and rendering vast areas of potential farmland unsuitable for cultivation or grazing.

In 1925 the cactus moth from Argentina was introduced to Australia to act as an agent of **biological control**. The larvae (see Figure 18.11) bored into the prickly pear cactus and spent many months living inside the plant and feeding on it. Once the plant was destroyed, the larvae pupated in the debris on the ground and later emerged as adult moths. This form of biological control was repeated generation after generation and eventually the prickly pear cactus in Australia was brought under control.

The same procedure was tried in other countries with mixed results. For example, in South Africa it proved to be problematic because the cactus moth larvae destroyed much of the population of spineless *Opuntia* that had traditionally been used as animal fodder. The cactus moth has now spread worldwide and is regarded as a pest by the ornamental cactus-growing industry in the USA.

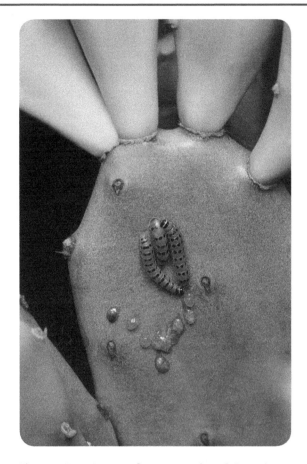

Figure 18.11 Larvae of cactus moth and *Opuntia*

Testing Your Knowledge

1 a) Briefly explain why, unlike other animals, the human population continues to increase rapidly. (2)
   b) Name TWO ways in which attempts are being made to increase food yield. (2)

2 a) Explain why fertiliser needs to be added to land where a crop is repeatedly grown, harvested and removed. (2)
   b) What events lead to the formation of an algal bloom? (2)
   c) Why does an algal bloom lead to a reduction in oxygen content of the water in the affected ecosystem? (2)

3 a) In general, what is the purpose of a pesticide? (1)
   b) Identify a possible adverse effect of the bioaccumulation of molecules of a non-biodegradable pesticide along a food chain. (1)

4 Define the term *biological control* and give ONE named example in your answer. (3)

19 Evolution of species

Mutation

A **mutation** is a change in the structure or composition of an organism's genetic material. If such a change in genotype produces a change in phenotype, the affected individual (such as the albino ivy plant shown in Figure 19.1) is called a **mutant**. In the absence of outside influences, mutations arise **spontaneously** and at **random** but only rarely.

Mutagenic agents

The mutation rate of a gene is expressed as the number of mutations that occur at that gene site per million gametes. The normal spontaneous rate of mutation can be increased by exposing cells to **mutagenic agents**. These are environmental factors such as some types of radiation and certain chemicals.

Figure 19.1 Albino ivy plant

Research Topic | **Mutagenic agents**

Mutagenic agents can be used by geneticists to try to create **new mutant varieties** of organism that are useful to humans. For example, the exposure of an enormous number of plants to X-rays occasionally produces an improved variety such as:

- a mutant strain of barley with an upright growth form that is resistant to heavy rain
- a mutant type of rice plant that shows improved hardiness to cold.

However, for the most part, mutagenic agents are **environmental hazards** that damage the genetic material of living things. In the experiment shown in Figure 19.2 an increased level of radiation brings about an increased rate of mutation but no new improved varieties are formed.

Mutagenic agents fall into three categories as follows.

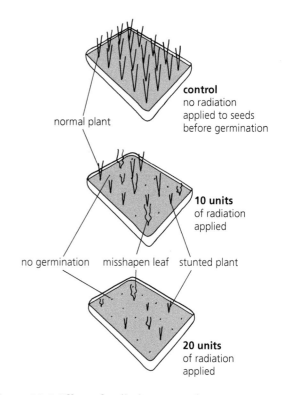

Figure 19.2 Effect of radiation on seeds

Ionising radiation

This group includes **X-rays** and radiation such as alpha, beta and gamma rays from **radioactive sources**. These rays damage genetic material directly by breaking it up. Occasional exposure of an individual's cells to X-rays for medical purposes is not harmful. However, the radiographer operating the equipment is at risk of frequent exposure and must be protected (for example, by a lead screen).

High levels of radiation from the fallout of a **nuclear bomb** or following an accident at a **nuclear power station** are seriously harmful or even lethal. The incidence of many forms of cancer (caused by mutations affecting stem cells) rose dramatically among the population living close to the Chernobyl nuclear power station in Ukraine following the disastrous accident that occurred there in 1986. Figure 19.3 shows one of the victims after thyroid cancer surgery.

Ultraviolet radiation

This form of radiation is also known as **UV light**. It is absorbed by some of the bases in DNA, which then become altered and linked together. Potentially this has a disruptive effect on the functioning of the DNA. However, in the absence of UV light, the linked bases are returned to normal by the action of **repair enzymes**.

Some people do not possess the correct allele of the gene that codes for one of the repair enzymes and may therefore develop **skin cancer** if exposed to high levels of ultraviolet radiation for prolonged periods. It is for this reason that people are advised to use a barrier cream with a high protection factor when sunbathing.

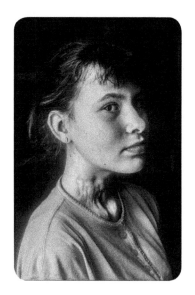

Figure 19.3 Victim of Chernobyl disaster

Chemical mutagens

Mustard gas is an example of a chemical mutagen. It acts by affecting one of the bases in DNA and making the molecule become unstable and ineffective. Mustard gas was first used as a chemical weapon in the First World War. It continues to be produced by several countries around the world and to be stockpiled as a means of defence in the event of war. When the gas comes into contact with humans, it acts as a powerful irritant that causes blistering of the skin, eyes and respiratory tract prior to damaging the DNA of the cells.

Mutation as a source of variation

Mutation is the only source of **new variation** among living things. It is the process by which new alleles of genes are produced. Without mutation, all organisms would be homozygous for all genes and no variation would exist.

Result of mutation

Most mutations are harmful (or even lethal) and therefore confer a **disadvantage** on the mutant organism's phenotype making it less likely to survive.

On rare occasions a mutant allele occurs by mutation that confers an **advantage** on the organism that receives it. The new allele results in the organism becoming better adapted to its environment. This increases its chance of survival. Such mutant alleles provide the alternative choices upon which **natural selection** can act (see page 166) and bring about evolutionary change.

The sequence of bases along a DNA strand contains genetic information that determines the sequence of amino acids in a protein (see Figure 3.4 on page 21). Mutations bring about changes to the genetic information held in the 'codewords' in the DNA.

Neutral mutation

A mutation may alter one base in the DNA strand and form a new codeword. The presence of this codeword introduces a chemically **similar amino acid** into the protein. The protein (and the mutant organism possessing it) remains unaffected. Such a mutation is described as **neutral**.

Disadvantageous mutation

If a DNA codeword is altered by mutation, this may lead to the **wrong amino acid** being introduced into the protein molecule at some point along its chain of amino acids or **no amino acid** being introduced at that point. This results in an inferior or non-functional version of the protein molecule being produced. For example, in humans a certain mutation on chromosome 7 leads to the formation of a protein that lacks one copy of an essential amino acid and is therefore non-functional. In the absence of the normal protein, the person secretes mucus that becomes thicker and stickier than normal. It causes congestion of organs such as the lungs and pancreas. This condition is called **cystic fibrosis** and a sufferer needs regular pounding on the chest to clear the thick mucus (see Figure 19.4).

Advantageous mutation

The most common form of the **peppered moth** is light brown in colour; a rarer mutant form is dark brown (almost black). This dark moth only differs from the light one by one allele of the gene controlling the formation of dark brown pigment. Both types of moth fly by night and rest on tree trunks during the day. In areas with clean air, many tree trunks bear pale-coloured lichens that provide camouflage for the light-coloured moths (see Figure 19.5), but not for the dark ones. These are easily spotted and eaten by predatory birds.

In areas suffering severe air pollution, toxic gases kill the lichens and soot particles darken the tree trunks. Now the situation is reversed. The **mutation** that resulted in the dark moth being almost black is **advantageous** since it promotes the dark moth's chance of survival in the polluted area.

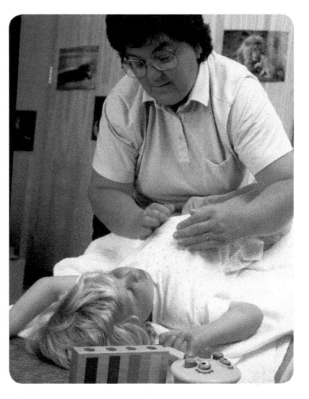

Figure 19.4 Treating a sufferer of cystic fibrosis

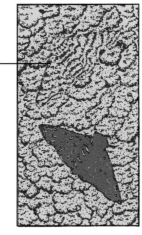
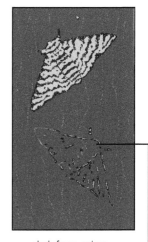

light form enjoys selective advantage on lichen covered trunk in non-polluted area

dark form enjoys selective advantage on soot covered trunk in polluted area

Figure 19.5 Advantageous mutation

Adaptation

An **adaptation** is an inherited characteristic that makes an organism well suited for survival in its environment. The members of a species possess certain **evolutionary adaptations**. These are the phenotypic expressions of changes that have occurred in the species' genotype as a result of mutations over a very long period of time. Possession of these adaptations gives the species some advantage and increases its chance of survival in its environment.

Related Activity

Investigating examples of adaptation

Desert plant

A normal land plant (such as an oak tree) would be unable to survive in the desert. Its leaves would present an enormous surface area of thin tissue from which water would be rapidly lost as water vapour by transpiration. At the same time, its roots would be unable to find water in the ground to replace this loss.

A desert plant such as a cactus is able to survive in such an extreme habitat because it possesses certain **adaptations**. Figure 19.6 shows the prickly pear plant. Its leaf surface is **greatly decreased** by its leaves being reduced to **protective spines**. Its stem is divided into green lobes that carry out photosynthesis. Each of these portions of stem is fleshy and stores water in **succulent tissues**. Water loss from the stem is kept to a minimum by the **thick, waxy cuticle** that coats its outer surfaces.

Some cacti have **long roots** for reaching supplies of subterranean water. Others possess extensive systems of **superficial roots** that grow parallel to the soil surface. These enable the plant to absorb maximum quantities of water on the rare occasions when rain does fall.

Galapagos finches

Charles Darwin visited the Galapagos Islands in 1835. He found them to be inhabited by many unique life forms including 13 different species of finch. Figure 19.7 shows a few of these finches. They are found to vary greatly in **beak size** and **shape**. These different **adaptations** make each species well suited to its environment by enabling it to exploit a particular ecological niche. For example, the insect-eating 'warbler' finch has a small, sharp, pointed beak that is ideal for picking insects out of narrow crevices in the bark of a tree. The seed-eating finch's large, blunt beak, on the other hand, is well suited to breaking open tough coats surrounding seeds and nuts.

Figure 19.6 Prickly pear plant

Figure 19.7 Darwin's finches

Plants and specific pollinators

Orchid and moth

The flowers of some plants are **adapted** to be pollinated by a specific animal that also possesses appropriate **adaptations**. For example, the star orchid (*Angreacum*) produces star-like flowers that release perfume at night. The back of the flower (see Figure 19.8) has a **long spur** with a **nectary** at the end of it. The plant's pollinator is a moth that flies by night and is attracted to the perfume. This type of moth has a **long proboscis** that enables it to reach the nectary. During feeding, it brings about pollination.

Figure 19.8 Star orchid

Such a specific relationship is beneficial to both the plant and the animal. The plant gets pollinated without its pollen being wasted on flowers of another species; the animal is guaranteed a food supply unavailable to other animal species.

Acacia and ant

Acacia is a plant commonly found as a shrub or tree in many parts of the world. One type of acacia native to Africa and South America is **adapted** in a way that enables it to benefit from a close relationship with its pollinator as follows.

The swollen-thorn acacia has **large, hollow thorns** that provide a habitat for a type of stinging ant (see Figure 19.9). The tree also supplies the ants with nectar. In return, the ants kill animals such as small herbivores that would attack the acacia plant and they destroy any plants growing close to the acacia that would compete with it for essential resources. The ants patrol 'their' acacia plant 24 hours a day and pollinate its flowers.

Figure 19.9 Swollen-thorn Acacia and acacia ants

Natural selection

Every species of living organism has an enormous **reproductive potential** (see Table 19.1). Its members produce far more offspring than the environment can support and this leads to a **struggle for survival**. A simplified example involving rabbits is shown in Figure 19.10. Many offspring die before reaching reproductive age as a result of factors such as:

- competition
- lack of food
- overcrowding
- inability to escape predators
- lack of resistance to disease.

Animal species	Mean no. of offspring per year
fox	5
red grouse	8
rabbit	24
mouse	30
trout	800
cod	4 000 000
oyster	16 000 000

Table 19.1 Reproductive potential

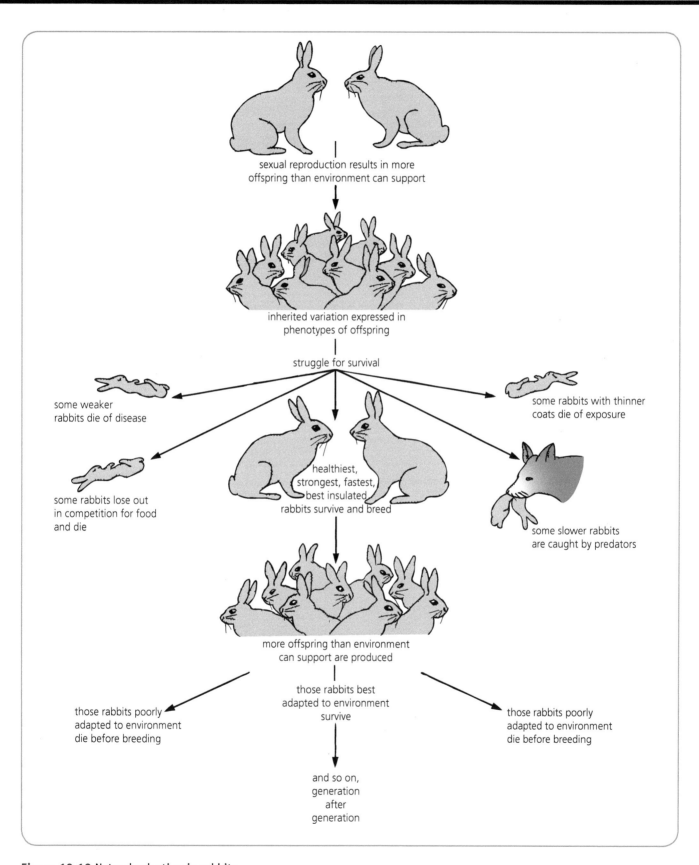

Figure 19.10 Natural selection in rabbits

The members of a species are not identical to one another but show phenotypic **variation** in all characteristics (see Figure 19.15 on page 170). Much of this variation is inherited. It is this inherited variation that is of significance in the process of natural selection.

Survival of the fittest

Those offspring whose phenotypes are **better adapted** to their environment have a better chance of surviving, reaching reproductive age and passing on the alleles for the favourable characteristics (that confer some advantage) to their offspring. These alleles increase in frequency within the population.

Those offspring whose phenotypes are least well suited to their environment die before reaching reproductive age and fail to pass on their alleles.

This process is repeated generation after generation and the organisms with the phenotypes best suited to the environment (i.e. the fittest) are 'selected' and predominate in the population. The poorer members are weeded out and die. They are the victims of **selection pressure**. The weeding-out process is known as **natural selection**. Charles Darwin first described natural selection in 1858. It is the main factor producing evolutionary change in species over time in response to changing environmental conditions.

Research Topic | Antibiotics and their over-prescription

Resistance to antibiotics

Antibiotics have been used in the UK for over 60 years. As each new antibiotic is introduced, it is found to be most effective against bacteria in its early years. Its effectiveness is found to decrease as the number of bacterial strains that are **resistant** to it increase and spread.

Micro-organisms such as bacteria occur in huge numbers. Genetic variation (resulting from spontaneous mutations) exists among the members of a population. A mutant bacterium may arise with genetic material that makes it resistant to a certain antibiotic without ever having been in contact with the antibiotic. Some disease-causing bacteria, for example, are able to make the enzyme penicillinase. This enables them to digest and resist penicillin. Figure 19.11 shows how resistant bacteria can spread from person to person.

Over-prescription of antibiotics

Sometimes antibiotics are prescribed for minor bacterial infections that, given time, would have been successfully dealt with by the body's natural defences alone. This **over-prescription** of antibiotics promotes an increase in the number and spread of bacterial strains resistant to the antibiotics. Antibiotics should only be used to treat bacterial infections that are so severe that they cannot be contained and destroyed by the body's own defences. This is often the case in young children and in the elderly.

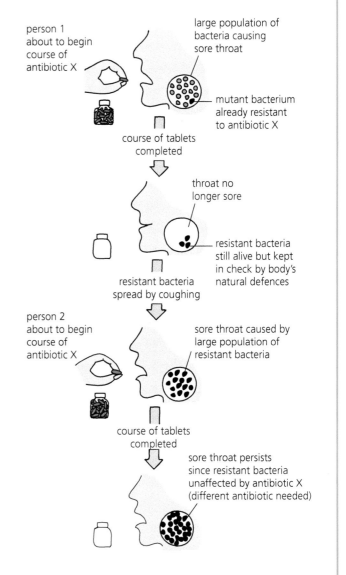

Figure 19.11 Spread of resistant bacteria

Research Topic | Rapid natural selection

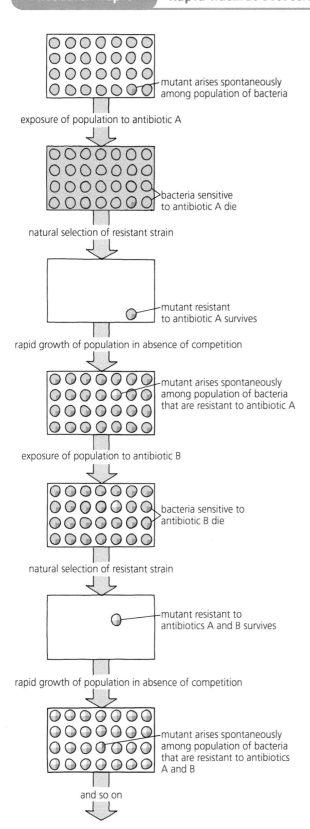

Figure 19.12 Evolution of resistant bacteria

The sequence of events that can lead to the evolution of mutant bacteria resistant to two different antibiotics is shown in Figure 19.12. As the process continues, some strains showing **multiple resistance** may continue to be favoured by this form of **rapid natural selection**. If these bacteria are also disease-causing, they will be difficult to treat.

Exchange of DNA

In addition to the transmission of genes for resistance to antibiotics from one generation of bacteria to the next, it is now known that portions of DNA including genes for antibiotic resistance can be passed from one type of bacterium to another type growing beside it in the same environment.

MRSA

Staphylococcus aureus is a species of bacterium commonly found on the surface of the human skin. It causes one in five of the infections acquired by hospital patients during treatment. In the past it was successfully controlled by a range of antibiotics. However, there now exists a strain of *S. aureus* known as **MRSA** (see Figure 19.13) that is **resistant** to almost all antibiotics.

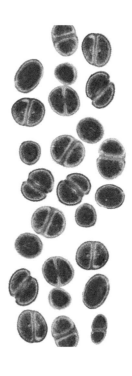

Figure 19.13 MRSA (methicillin-resistant *Staphylococcus aureus*)

Case Study | Insect resistance to 'built-in' insecticides in GM crops

Pink bollworm

The pink bollworm (see Figure 19.14) is the larval stage of a type of moth. The female adult lays her eggs in a cotton boll. When the eggs hatch, the larvae that emerge feed on and damage the cotton plant. Genetic engineering has been used to produce a transgenic variety of cotton plant containing a gene from a bacterium that makes an insecticide lethal to bollworms. This chemical gives these genetically modified (GM) cotton plants protection against the bollworm pests. However, in some parts of India, a strain of pink bollworm has already been discovered that is resistant to the 'built-in' insecticide present in the GM cotton crop.

European corn borer

The corn borer is also the larval stage of a type of moth. The female lays her eggs on the underside of the leaves of maize ('corn-on-the-cob') plants. When the caterpillars (borers) hatch from their eggs, they damage the cobs and tunnel into the stalks, making the plants fall over. A transgenic variety of maize has been produced that contains a bacterial gene for an insecticide. This toxic chemical is lethal to corn borers, enabling the GM maize crop to protect itself against these pests.

However, it is only a matter of time until a strain of corn borer resistant to the 'built-in' insecticide appears. Already scientists have produced a population of resistant corn borers by repeatedly exposing colonies of the insect to the insecticide in the lab. To try to delay the natural emergence of resistant corn borers on farmland, farmers in the USA who plant GM maize are obliged by law to plant normal, non-transgenic maize nearby. These plants are intended to provide a location to harbour the pests and slow down the evolution of the type resistant to the 'built-in' insecticide.

Figure 19.14 Pink bollworm

Figure 19.15 Variation within a population of limpets

Speciation

A **species** is a group of living things that are so similar to one another genetically that they are able to interbreed and produce **fertile** offspring.

Speciation is the formation of **new** species. It occurs when a population becomes **isolated** from the other members of its species for a very long time or permanently. The isolated population finds itself subjected to environmental conditions different from the ones to which it was previously adapted. Under these circumstances, natural selection may take a **new direction** and result in the isolated population eventually becoming a new species.

The simplified version of speciation shown in Figure 19.16 occurs as follows.

1 The members of a large population of a species occupy an environment and **interbreed,** forming fertile offspring.

2 The population becomes split into two completely isolated populations, X and Y, by a **barrier** that prevents interbreeding and exchange of genetic material. This isolating mechanism could, for example, be

- **geographical** in the form of a river, a desert, a sea or a mountain range
- **ecological** caused by differences in temperature, humidity, pH or salinity.
- **behavioural** involving exploitation of different types of food.

3 **Mutations** occur at random. Most of the mutations affecting population X are different from those affecting population Y. Any new variation that arises in one group is not shared by the other group. For example, the small mutant that arises in group Y does not occur in group X.

4 The **selection pressures** acting on each population are different depending on environmental conditions such as type of climate, predators and disease. In this example the isolating barrier has resulted in population Y landing in a region with a warmer climate than the one affecting population X. **Natural selection** affects each population in a different way by favouring those alleles that make the members of that population best at exploiting their environment. For example, small-sized members of population Y will be selected because they are better suited to survival in a warm climate than larger members.

5 Over a very long period of time, stages 3 and 4 cause the two groups to become **genetically distinct**.

6 If the barrier is removed, they are no longer able to interbreed successfully. **Two distinct species** now exist: species X (similar to the original species) and species Y (a completely new species).

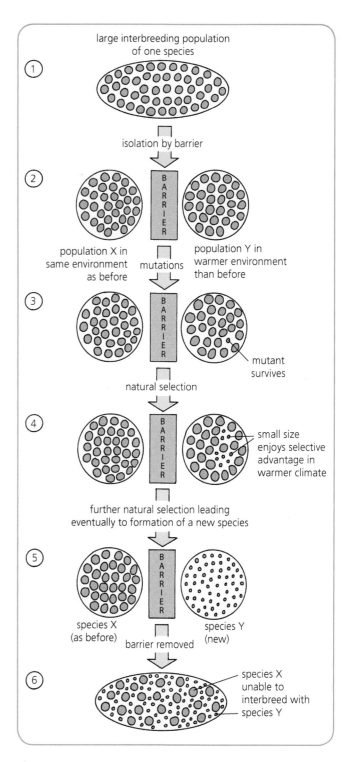

Figure 19.16 Speciation

| **Isolation leading to speciation**

Endemic species

When a particular type of organism is found only in a certain region such as an island, it is said to be **endemic** to that region. Endemic species of plants and animals are found to occur on some Scottish islands.

Arran whitebeam

The **mountain ash** is a type of tree commonly found throughout Europe, especially on rocky mountainous soil but not on clay or limestone. The **rock whitebeam** is a shrub (or rarely a small tree) found on crags among rocks, especially on limestone soil. The **Arran whitebeam** is a small, slender tree, endemic to the island of Arran, where it grows on steep granite banks and mountain slopes. It favours high-altitude environments near the tree line where competition is reduced.

The Arran whitebeam is believed to have arisen as a result of hybridisation between the mountain ash and the rock whitebeam. By growing in habitats different from those favoured by its parents and lacking competitors such as those found on the mainland, the Arran whitebeam has taken its own course of evolution and is a different species from both of its ancestors (see Figure 19.17).

St Kilda wren

St Kilda is a group of tiny islands in the Atlantic Ocean isolated from the Scottish mainland (see Figure 19.18). The type of wren (the **St Kilda wren**) that is endemic to this island environment differs from its mainland relative in several ways. The St Kilda wren:

- is heavier
- has longer wings
- has thicker legs
- is paler and more striped
- lays larger, heavier eggs

mountain ash

rock whitebeam

Arran whitebeam

Figure 19.17 Arran whitebeam and relatives

It is possible that these differences are **adaptive features** that help the bird to survive among the rocks in such a wet and windy environment totally lacking in trees. Although **speciation** is well under way, the St Kilda wren has not been isolated from the mainland population for long enough yet to have become a separate species. It is still able to mate with the mainland variety and produce fertile offspring.

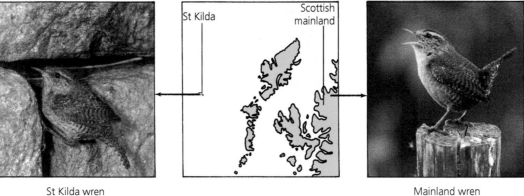

St Kilda wren | St Kilda | Scottish mainland | Mainland wren

Figure 19.18 Two types of wren

Testing Your Knowledge

1 a) What is meant by the term *mutation*? (1)
  b) What is a *mutant*? (1)
  c) i) Give an example of a mutagenic agent.
    ii) What effect does a mutagenic agent have on an organism's natural mutation rate? (2)

2 a) Table 19.1 on page 166 shows that an oyster has the potential to produce 16 million offspring annually. Give THREE reasons why the seas are not over-populated with oysters. (3)
  b) In the struggle for survival, which oysters will:
    i) survive and pass on their genes to succeeding generations?
    ii) die before reaching reproductive age? (2)
  c) What is the name given to this 'weeding out' process that promotes survival of the fittest? (1)

3 a) Explain the difference between the terms *species* and *speciation*. (2)
  b) Copy and complete the paragraph below using the following terms: *advantage, isolated, mutations, natural selection, speciation*. (4)

    If a population of organisms becomes _____ from the other members of its species, it may undergo

_____ and eventually become a new species. This happens if _____ follows a different path in this new environment and some _____ that would have been neutral or disadvantageous before now confer an _____ on the organisms.

4 Decide whether each of the following statements is true or false and use T or F to indicate your choice. Where a statement is false, give the word(s) that should have been used in place of the word(s) in **bold** print. (5)
  a) **Adaptation** is the only source of new variation among living things.
  b) Most mutations confer **an advantage** on the mutant organism.
  c) Some mutations are **neutral** and have no effect on the mutant's phenotype.
  d) An inherited characteristic that makes an organism well suited to its environment is called an **evolution**.
  e) **Natural selection** is the process by which the members of a population best adapted to changing environmental conditions survive and pass their genes on to succeeding generations.

What You Should Know Chapters 18–19

adaptation	fertiliser	offspring
adapted	fittest	oxygen
advantage	genetic	pesticides
alleles	increase	protein
amino	intensive	radiation
bacteria	isolation	random
biological	mutagenic	selection
bloom	mutation	speciation
disadvantage	natural	toxicity
enemies	nitrate	yield
fatally	non-biodegradable	

1 The human population continues to _____ rapidly and this needs to be matched by an increase in food _____, which often involves the use of fertilisers and _____.

2 _____ dissolved in soil water is absorbed by plants and used to make _____ acids. These are built into plant protein. On being consumed by animals, plant _____ provides amino acids for the synthesis of animal protein.

3 When _____ from farmland leaches into watercourses or lochs it may over-enrich the water and cause formation of an algal _____. When the algae die, the decomposer _____ that act on them undergo a population explosion and use up the water's dissolved _____ supply during respiration.

4 Molecules of pesticides that are _____ can accumulate in the bodies of organisms. This leads to an increase of _____ along food chains which may affect the final consumers _____.

5 Attempts are made to balance the adverse effects of _____ farming by using GM crops and employing _____ control as alternatives to the use of

pesticides. Biological control is the reduction of a pest population by the introduction of one of its natural _____.

6 A _____ is a change in the structure or composition of an organism's _____ material. It may confer an advantage or a _____ on the organism, or be neutral and not affect it.

7 Mutations occur spontaneously and at _____, but rarely. They are the only source of new _____.

8 The rate of mutation can be increased by _____ agents such as _____ and certain chemicals.

9 An inherited characteristic that makes an organism well suited to survive in its environment is called an _____.

10 The members of a species produce far more _____ than the environment can support. This leads to a struggle where only the _____ survive.

11 Those individuals best _____ to a changing environment, pass on the genes that confer a selective _____ to their offspring. The less well adapted members die. This weeding-out process is called _____ selection.

12 _____ occurs when a population becomes isolated for a very long time and natural _____ takes a new direction. Gradually the group becomes adapted to environmental conditions that differ from those that affected it before _____

Applying Your Knowledge and Skills Chapters 14–19

1 Figure KS3.1 shows a simple food web.

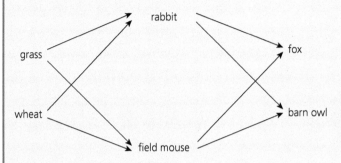

Figure KS3.1

Decide whether each of the following statements is true or false and use T or F to indicate your choice. Where a statement is false, give the word(s) that should have been used in place of the word(s) in **bold print**.

a) Each arrow represents the flow of **energy** from one organism to the next. (1)
b) The primary consumers are **larger** in body size than the secondary consumers. (1)
c) Foxes and barn owls are the **rarest** organisms in this food web. (1)
d) Both field mice and **foxes** are examples of primary consumers. (1)
e) The total quantity of energy that passes from plants to rabbits is **less** than the total quantity of energy that passes from rabbits to foxes. (1)
f) The producers are eaten by the **secondary** consumers. (1)

2 The information in Table KS3.1 was collected by studying eight different species of snail present in six Scottish lochs. (+ = snails present) (ppm = parts per million)
a) What concentration of calcium was present in loch 5? (1)
b) Which loch contained 9.5 ppm of calcium in its water? (1)
c) Which species of snail was found in every loch? (1)
d) Which loch possessed only five different species of snail? (1)
e) How many different species of snail were found in loch 2? (1)

f) i) What relationship seems to exist between the number of species of snail present and the concentration of calcium in the water?
   ii) Give a possible explanation for this relationship based on the fact that one of the components of a snail's shell is calcium. (2)

| Loch | Species of snail | | | | | | | | Concentration of calcium (ppm) in loch water |
	A	B	C	D	E	F	G	H	
1	+	+		+	+	+	+	+	19.9
2		+	+		+	+	+	+	17.8
3			+	+	+		+	+	15.3
4				+		+	+	+	9.5
5							+	+	7.1
6								+	5.2

Table KS3.1

3 a) Copy and complete Table KS3.2, which refers to some of the design features that apply to the cress seed investigation on pages 122–123. (4)

Feature of experimental design	Reason
equal mass of cotton wool and equal volume of water added to each carton	
yoghurt cartons are encased in black paper or painted black	
only recently purchased cress seeds are used	
experiment is repeated many times and results are pooled	

Table KS3.2

b) It could be argued that the water was lost by the evaporation from the two cartons at different rates and that this introduced a second variable factor. Suggest an improvement that could be made to the design to overcome this problem. (1)

4 Figure KS3.2 shows the results from an investigation into the numbers of two closely related species of spider, X and Y, living on a climbing ivy plant. The size of the circle at each sample site indicates the relative size of the total population of spiders found there. The black portion indicates species X and the blue portion species Y.

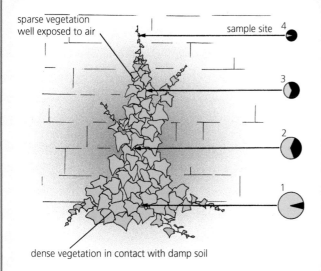

sparse vegetation well exposed to air

sample site 4

dense vegetation in contact with damp soil

Figure KS3.2

a) What relationship exists between the *total* number of spiders and the distance from the ground? (1)
b) Which species is more numerous at ground level? (1)
c) Describe the trend in relative numbers of the two species of spider that occurs from site 1 to site 4. (1)
d) i) Predict what will happen when equal numbers of the two species of spider are placed in the plastic tube shown in Figure KS3.3.

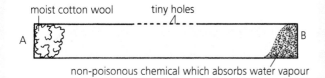

moist cotton wool   tiny holes

A   B

non-poisonous chemical which absorbs water vapour

Figure KS3.3

ii) Explain your answer. (2)

5 The information in Figure KS3.4 and the key of paired statements below refer to eight different fish.

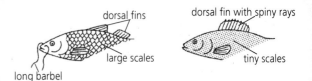

dorsal fins

dorsal fin with spiny rays

large scales

tiny scales

long barbel

Figure KS3.4

KEY

1

body with large scales .... go to **2**

body with tiny scales.................................... go to **5**

2

one dorsal fin present.................................... go to **3**

two dorsal fins present, the first with
spiny rays .... go to **4**

3

dark spot behind head .... **shad**

dorsal fin placed far back on body .... **pike**

4

dark spots on first dorsal fin forming bands...... **zander**

dark spot on rear end of first dorsal fin........... **perch**

5

two or more barbels around the mouth..... go to **6**

no barbels present .... go to **7**

6

one long barbel on lower jaw, two barbels
at nostrils... **burbot**

two long barbels on upper lip .... **catfish**

7

one long dorsal fin present..................... **blenny**

two dorsal fins present, the first with
spiny rays .... **miller's thumb**

a) Identify fishes W and X shown in Figure KS3.5. (2)

fish W   fish X

Figure KS3.5

b) Give TWO characteristics of a burbot. (2)

c) Name TWO features shared by a zander and a perch. (2)

d) Identify fishes Y and Z from the following descriptions:

  i) Fish Y has two dorsal fins, the first with spiny rays. It has a fairly broad head with the eyes on the top of it. It is a small fish and has neither barbels nor large scales.

  ii) Fish Z lacks a lateral line but does have one black spot behind its head and one dorsal fin. It has large scales and those on its belly form a sharp toothed edge. (2)

e) State TWO differences between a pike and a miller's thumb. (2)

6 Figure KS3.6 shows a form of atmospheric pollution. Match the numbered boxes in the diagram with the following lettered descriptions to show the correct sequence of events. (1)

Figure KS3.6

A acid rain falls on land and water environments

B fossil fuels burned by industry, power stations and motor vehicles

C acid gases form acid rain clouds

D trees suffer dieback and fish die

E acid gases released into the atmosphere

7 The graph in Figure KS3.7 shows the results from two experiments set up to investigate the rate of photosynthesis of a water plant exposed to different environmental conditions. The CO_2 concentration of the pond water was maintained at a very high level throughout the two experiments.

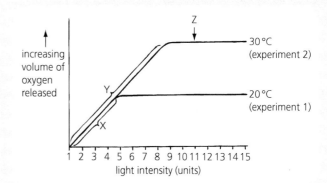

Figure KS3.7

a) In experiment 1, the temperature was kept constant at 20 °C.

  i) Name the environmental factor that was varied.

  ii) Name a second environmental factor that was kept constant. (2)

b) In the second experiment, the temperature was kept constant at 30 °C. Suggest how this was done. (1)

c) What factor was limiting the rate of photosynthesis at region X in the graph? (1)

d) State the light intensity at which a temperature of 20 °C first began to limit the photosynthetic rate. (1)

e) State the factor that limited the photosynthetic rate:

  i) in region Y

  ii) at point Z on the graph for experiment 2. (2)

8 Figure KS3.8 shows the number of units of energy (in kilojoules/m²/year) that are transferred from organism to organism in a pond food chain.

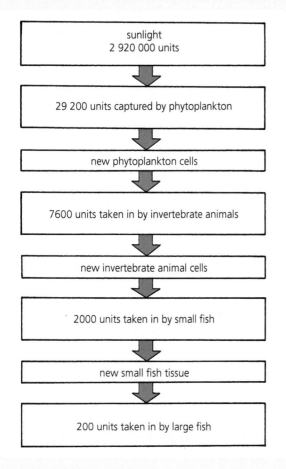

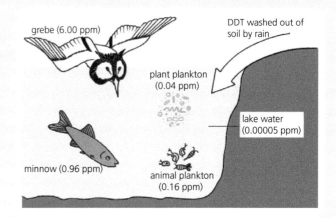

Figure KS3.8

a) What percentage of sunlight is successfully captured by the phytoplankton? (1)
b) What percentage of energy is lost between intake of energy by small fish and intake of energy by large fish? (1)
c) How many units of energy are lost when small fish feed on invertebrate animals? (1)

9 Figure KS3.9 shows the concentration of the pesticide DDT in parts per million (ppm) found in the cells of the members of a loch community.

Figure KS3.9

a) Construct a food chain that includes the four types of organism shown in the diagram. (1)
b) Construct a table to show the concentration of DDT found in the cells of each type of organism with the information arranged in increasing order. (2)
c) i) By how many times did the concentration of pesticide increase between loch water and plant plankton?
   ii) Between which two organisms did the pesticide concentration increase by exactly six times? (2)
d) i) Which animal do you think was found to suffer most and often die as a result of DDT poisoning?
   ii) Explain your choice of answer. (2)
e) Suggest why the use of DDT is now banned in Britain. (1)

10 The graph in Figure KS3.10 refers to an investigation into the effect of increasing radiation on the percentage number of X chromosomes showing a lethal (deadly) mutation. The animal used was the fruit fly (*Drosophila melanogaster*). The results are plotted as points with the line of best fit drawn through them.

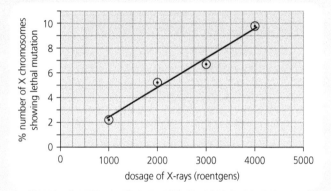

Figure KS3.10

a) What was the one variable factor altered by the experimenter in this investigation? (1)
b) How many conditions of this factor were used? (1)
c) Name TWO other factors that would have to be kept constant to make the experiment valid. (2)
d) What conclusion can be drawn from the above results? (1)
e) Explain why it is acceptable that a hospital patient with a suspected fractured limb be exposed to X-rays yet the radiographer must remain behind a lead screen to avoid exposure. (1)

11 Figure KS3.11 refers to three subspecies of the European wren.

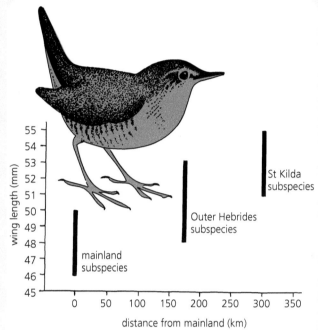

Figure KS3.11

a) What is the range of wing lengths found amongst the St Kilda wrens? (1)
b) How many of the three subspecies contain birds with a wing length of:
   i) 47 mm?
   ii) 49 mm?
   iii) 52 mm?
   iv) 54 mm? (4)
c) What relationship exists between wing length and distance of the bird's habitat from the mainland? (1)
d) Identify the barrier that isolates the three subspecies of wren. (1)
e) By what means could scientists find out if the three subspecies have become distinct species? (1)

(Since this group of questions does not include examples of every type of question found in SQA exams, it is recommended that students also make use of past exam papers to aid learning and revision.)

Appendix 1

Units of measurement

Quantity	Standard unit	Symbol	Relationship between units
length	**metre**	m	
	other useful units:		
	centimetre	cm	$1\,cm = 1/100th$ of a metre $= 10^{-2}\,m$
	millimetre	mm	$1\,mm = 1/1000th$ of a metre $= 10^{-3}\,m$
	micrometre	µm	$1\,µm = 1/1000th$ of a millimetre $= 10^{-3}\,mm$
			$1\,µm = 1/1000\,000th$ of a metre $= 10^{-6}\,m$
area	**square metre**	m^2	
volume	**cubic metre**	m^3	
	other useful units:		
	cubic decimetre (litre)	dm^3 (l)	$1\,dm^3 = 1/1000th$ of a cubic metre
	cubic centimetre	cm^3	$1\,cm^3 = 1/1000th$ of a cubic decimetre (litre)
mass	**kilogram**	kg	
	other useful units:		
	gram	g	$1\,g = 1/1000th$ of a kilogram
	tonne	t	$1\,t = 1000\,kg$
energy	**joule**	J	
	other useful unit:		
	kilojoule	kJ	$1\,kJ = 1000\,J$
time	**second**	s	
	other useful units:		
	minute	min	$1\,min = 60\,s$
	hour	h	$1\,h = 60\,min = 3600\,s$
temperature	**degree Celsius**	°C	

Table Ap 1.1

Appendix 2

pH scale

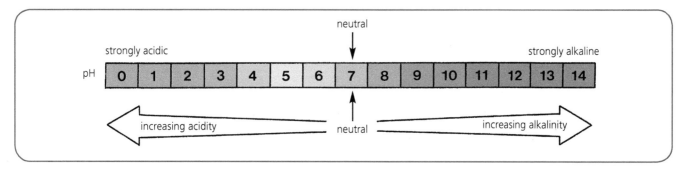

Figure Ap 2.1

Appendix 3

Size scale of plants and micro-organisms

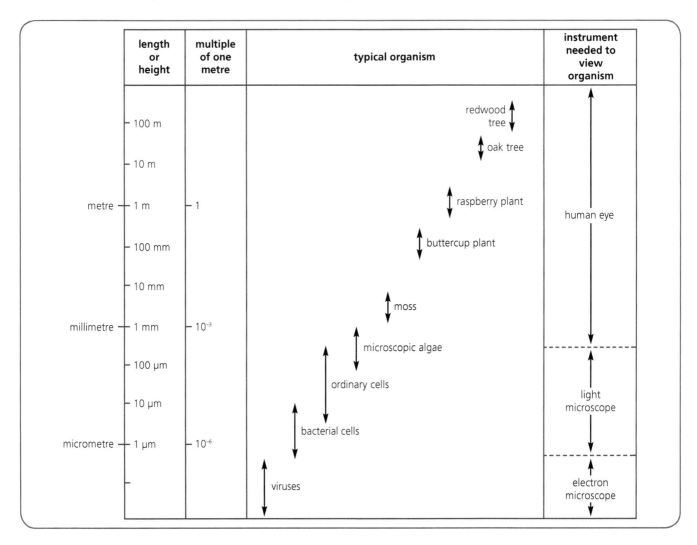

Figure Ap 3.1

Variable factors

Each of the pots shown in Figure Ap 4.1 contains the same **number** of seeds planted at the same **soil depth**. These words in bold print refer to factors that are constant in Figure Ap 4.1 but that could be varied. These are therefore called **variable factors**.

In a scientific investigation, a test is **valid** if it deals with **one variable factor** at a time, as shown in Figures Ap 4.2 and Ap 4.3. However, if the parts of an experiment differ from each other by more than one variable factor, the test is unfair and therefore invalid. In Figure Ap 4.4, the pots differ from one another in two ways:

- seed number
- depth of planting.

This is an invalid test because it will be impossible to say whether the seeds that grow best have done so due to their number or their depth of planting.

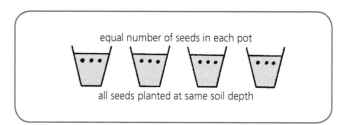

Figure Ap 4.1 Variable factors kept constant

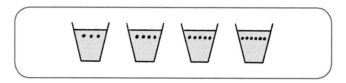

Figure Ap 4.2 Seed number as variable factor

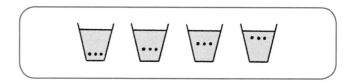

Figure Ap 4.3 Depth of planting as variable factor

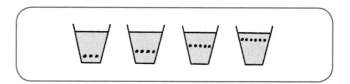

Figure Ap 4.4 An invalid test

Index